SpringerBriefs in Earth Sciences

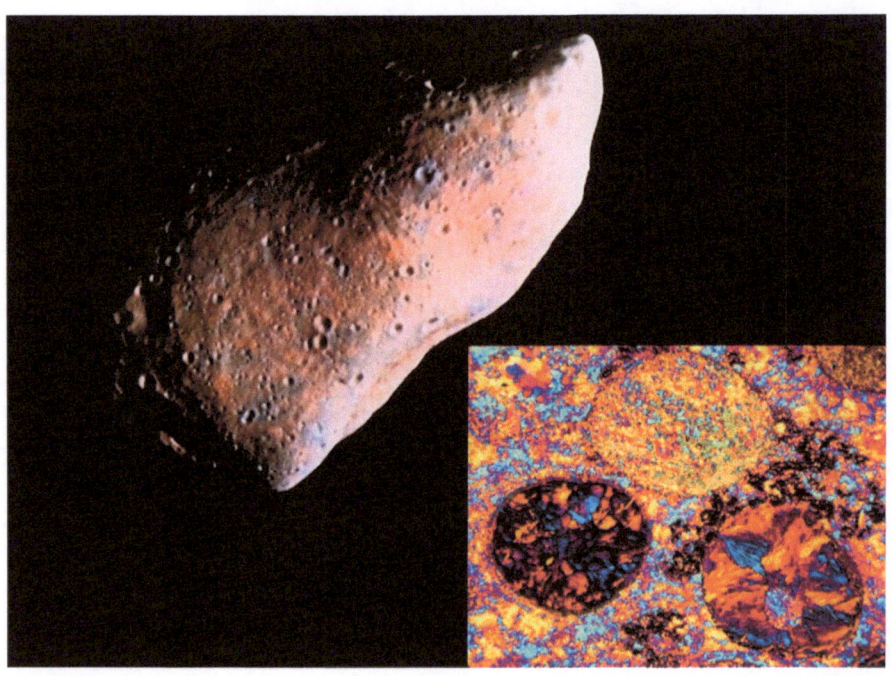

The asteroid Eros (34.4 × 11.2 × 11.2 km) (NASA) and S3 impact microkrystites, Barberton Greenstone Belt (courtesy G. Byerly)

Andrew Y. Glikson

The Asteroid Impact Connection of Planetary Evolution

With Special Reference to Large Precambrian and Australian Impacts

 Springer

Andrew Y. Glikson
School of Archaeology and Anthropology
Australian National University
Canberra, ACT
Australia

ISSN 2191-5369   ISSN 2191-5377   (electronic)
ISBN 978-94-007-6327-2   ISBN 978-94-007-6328-9   (eBook)
DOI 10.1007/978-94-007-6328-9
Springer Dordrecht Heidelberg New York London

Library of Congress Control Number: 2013931217

Printed on acid-free paper

Springer is part of Springer Science+Business Media (www.springer.com)

*In honor of Eugene M. Shoemaker,
Carolyn S. Shoemaker, Robert S. Dietz,
David H. Green and Ian Williams*

Preface

A paradigm shift according to Thomas Kuhn (1962) constitutes a change in the basic assumptions within the ruling theory of science. It is not a term to be used lightly, except in relation to major breakthrough in the understanding of nature. In the field of Earth Science this term can be used in connection with the conception of gradualism in terrestrial evolution by James Hutton (1788) and Charles Lyell's (1830), sea floor spreading and plate tectonics by Harry Hess, Bruce Heezen, Robert Dietz, and Sam Carey, and the identification of meteorite craters and astroblemes ('star scars') by Eugene Shoemaker and Robert Dietz, both having been my mentors. My introduction to extraterrestrial impacts in 1968 was related to the study of Gosses Bluff Structure, Central Australia, where the United States Astrogeology Branch, led by Eugene Shoemaker, was planning a study of Moon-like landscapes in preparation for the Apollo program (Fig. 1.1). At the time few geologists realized the role of asteroid impacts. In subsequent years, the sea-change discovery by Walter and Louis Alvarez of the KT asteroid impact boundary and associated mass extinction of species has changed this attitude. This was followed by the identification of the relations between the 580 Ma-old Acraman impact structure, the Bunyeroo ejecta, and radiation of Acritarchs by George Williams, Victor Gostin, and Kath Grey. Based on geological studies of Archaean terrains during the 1980s and 1990s I raised doubts whether many Precambrian Earth features were triggered exclusively by internal mantle and crust processes. A breakthrough came in 1986 and following years when Don Lowe, Gary Byerly, Bruce Simonson, and Scott Hassler and their students began to discover millimeter scale impact spherules (microkrystites) in Archaean sediments, overlain by tsunami deposits, initiating a paradigm shift in the study of early crustal evolution. Given the difficulty in identifying spherule units in the field, impact frequencies documented to date inherently represent only a minimum flux, namely the 'tip of the iceberg', yielding support to an extension of the Late Heavy Bombardment. This monograph, focusing on impacts craters larger than 20 km in diameter, is based on research of Archaean and younger terrains during 1964–2012, including studies of impact ejecta units and large buried impact structures on the Australian continent. Notably detailed research in the Pilbara Craton, with the support of Arthur Hickman of the Western Australian Geological Survey and my field mate John Vickers, enabled follow-up of discoveries by Lowe, Byerly, Simonson and their students. Suggestions that Archaean extra-terrestrial impacts acted as triggers of internal mantle-crust events will meet with resistance by proponents of uniformitarian schools of thought. Traditionally, geology—the study of Earth—focuses on internal crust, mantle, and core process, taking little account of the effects of large asteroid impacts. However, the two are not mutually exclusive. Whereas purely endogenic mantle-crust dynamics and plate tectonic cycles are manifest, the intermittent triggering of thermodynamic events by large extra-terrestrial impact clusters constitutes a combination of Cuvier's catastrophism and Lyell and Hutton's gradualism throughout Earth history.

Reference

Kuhn TS (1962) The structure of scientific revolutions. The University of Chicago Press, Chicago

Acknowledgments

This book is based to a large extent on my field and laboratory investigations of impact ejecta units in the Pilbara, Western Australia, on laboratory studies of impact ejecta from the Barberton Greenstone Belt, South Africa, during 1998–2007, and on studies of Australian buried impact structures during 1999–2012. The book was invited by Petra Van Steenbergen and edited by Hermine Vloemans of Springer SBM NL. I am grateful to Don Lowe, Gary Byerly, Franco Pirajno, Victor Gostin, Hugh Davies, Miryam Glikson, Peter Haines, and Arthur Hickman for their comments on the book manuscript. I am indebted to John Vickers, my field mechanic and laboratory technician, for consistent help and geological interest in our investigations. Field work in the Pilbara would not have been possible without the long-term support and interest of Arthur Hickman of the Geological Survey of Western Australia. I thank Robert Iasky, Franco Pirajno, Peter Haines, Martin Van Kranendonk, and John Gorter for their interest and numerous discussions. I am grateful to Bruce Simonson and Scott Hassler for introducing me to impact ejecta exposures in the Hamersley Range, to Gary Byerly and Don Lowe for providing ejecta samples from the Barberton Greenstone Belt. Tonguc Uysal and Hal Gurgenci collaborated in the study of the Warburton shock metamorphic terrain. David Green and Ian Jackson facilitated my research at the Research School of Earth Science, Australian National University. I thank Charlotte Allen, Harry Kokkonen, Frank Brink, Stephen Eggins, John Fitz Gerald, and Tony Eggleton for help and advice with analytical work. I thank David Jablonski for collaboration with the Mount Ashmore study. I thank Alan Whittaker, Elinor Alexander, Rodney Boucher, John Bunting, Prame Chopra, Chris Klootwijk, Nick Lemon, Tony Meixner, Martin Norvick, Hugh O'Neill, Bruce Radke, Erdinc Saygin, John Veevers, Xiaowen Sun, and Doone Wyborn for discussions regarding the Warburton structure, and Les Tucker, David Groom, Karen Groom, and Michael Willison of PIRSA for help with examination and sampling of drill cores, Elaine Appelbee for drafting. I am grateful to the following people for permission to use figures in this book: Alessandro Montanari, Alex Shukolyukov, Anita Andrews, Karen Ballen, Gary Byerly, Sherry Cady, Bevan French, Richard Grieve, Duane Hamacher, Scott Hassler, Arthur Hickman, Alan Hildebrand, Michael Jones, Gerta Keller, Don Lowe, Victor Masaitis, Mustafa Mincel, Victor Gostin, Reg Morrison, Franco Pirajno, Bruce Simonson, Caroline Shoemaker, John Spray, Claudia Trepmann, Martin Van Kranendonk, and Jim Wark.

Contents

Chapter 1
A Paradigm Shift in Earth Science

Abstract This section suggests the evolution of Earth progressed through a combination of internal core-mantle-crust dynamics and extraterrestrial large impacts which triggered seismic, tectonic, volcanic and tsunami processes as well several mass extinctions of species.

When, in 1981, Louis and Walter Alvarez, the father and son team, unearthed a tell-tale Iridium-rich sedimentary horizon at the ~65 million years-old (Ma) Cretaceous-Tertiary (KT) boundary at Gubbio, Italy (Alvarez et al. 1980, 1982; Alvarez 1997), the find heralded a paradigm shift in the study of terrestrial evolution. The discovery re-established the idea that much of Earth's history has been shaped by catastrophes, a theory promoted by Georges Cuvier and natural theologians which preceded, but was largely supplanted by, Darwin's (1859) theory of evolution and by Hutton (1788) and Lyell's (1830) geological gradualism (Fig. 1.1).

The KT boundary (64.98 ± 0.05 Ma[1]) corresponds to the 2nd largest mass extinction of species recorded in Earth history, when some 46 % of living genera were extinguished (Keller 2005) (Figs. 1.2, 9.3 and 9.4). Since the parent craters of the KT event have been identified, including *Chicxulub* (170 km-diameter, Yucatan Peninsula, Mexico) and *Boltysh* (~25 km-diameter, 65.17 ± 0.64 Ma, Ukraine), other large asteroid impact craters and impact ejecta units have been associated with mass extinction and biological radiation boundaries (Grey et al. 2003; Grey 2005; Glikson 2005, 2009), underpinning the vulnerability of species to catastrophic events.

The Earth conceals its secrets well, not least buried scars of meteorite impacts and thin, commonly hardly detectable, spherule layers within sedimentary sequences. Based on shock metamorphic criteria calibrated with laboratory experiments (French 1998), the pressure–temperature field of shock metamorphism is distinct from that of terrestrial metamorphism, including that of high pressure eclogite facies (Fig. 1.3). Whereas the PT field of eclogite is below 10 GPa, the graphite to diamond and coesite to stishovite transformations, shatter cones, planar deformation features, diaplectic glass and shock melting occur well above 10 GPa (Fig. 1.3).

---

[1] Impact structure ages and diameters are after the Earth Impact Database [EID] (http://www.passc.net/EarthImpactDatabase/index.html) and information by the author. Where two figures are cited (cf. 130 < 260 km) the lower value represents an estimate of the diameter of the collapsed crater whereas the higher value is the diameter of the outer ring.

A. Y. Glikson, *The Asteroid Impact Connection of Planetary Evolution*,
SpringerBriefs in Earth Sciences, DOI: 10.1007/978-94-007-6328-9_1,
© The Author(s) 2013

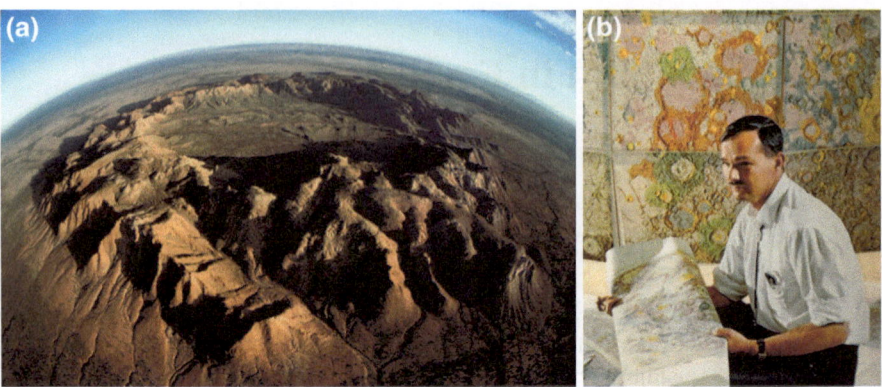

Fig. 1.1 Gosses Bluff impact structure, central Australia, and Eugene Merle Shoemaker compiling the first comprehensive geologic moon map http://users.tpg.com.au/users/tps-seti/planets.html. In 1968 the joint Gosses Bluff study by the U.S. Geological Survey, headed by Eugene Shoemaker, and the Australian Bureau of Mineral Resources, opened my eyes to the significance of large asteroid impacts and their diagnostic hallmarks, to be followed years alter by systematic studies of Archaean impactites and buried impact structures on the Australian continent. **a** Aerial view of Gosses Bluff, looking from the south (courtesy Reg Morrison); **b** Eugene Shoemaker studying lunar maps (courtesy Carolyn Shoemaker)

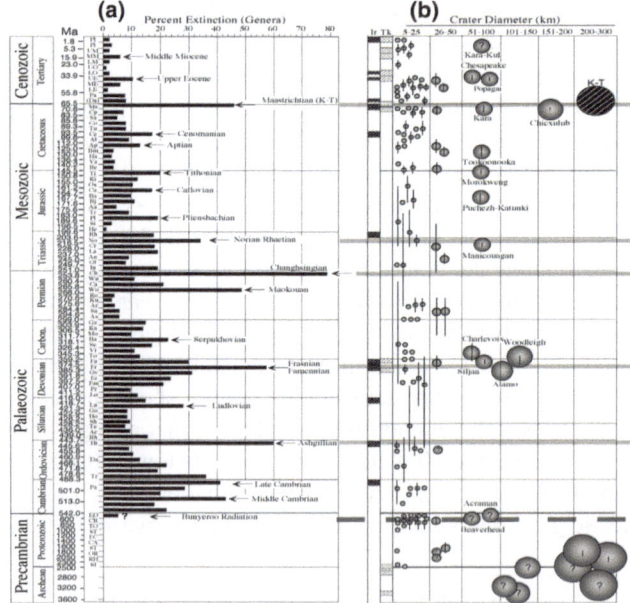

Fig. 1.2 Phanerozoic mass extinctions, asteroid impacts, and large igneous provinces. **a** Extinction intensity; **b** Impact events; **c** Volcanism. Stratigraphic subdivisions and numerical ages are after Gradstein et al. 2004). The extinction record is based on genus-level data by Sepkoski (1996). The number of impact events, size and age of craters follows largely the Earth Impact Database (2005), with modification by the author (AG). (Keller 2005, by permission)

Where signatures of buried extraterrestrial impact structures are found, their structure and composition needs testing by geophysical and geochemical methods, while confirmation often has to wait for years before their origin can be established. Until the 1960s, the apparent scarcity of large impact basins on Earth, as contrasted with the large lunar mare basins, constituted a major objection to theories advocating catastrophic extraterrestrial collisions. Several scientists, including Alt et al. (1988), Oberbeck et al. (1992) and Abbott and Isley (2002), proposed genetic connections between large impacts and geodynamic events. Following the establishment of criteria for shock metamorphism at Meteor Crater, Arizona (Shoemaker and Kieffer 1979; Roddy and Shoemaker 1995) and the Ries crater, Germany (cf. Chao 1967, 1968), the pioneering studies by Robert Dietz have paved the way for identification of giant impact structures, referred to as *astroblemes* (star scars). This included Vredefort (289 km-diameter; 2,023 ± 4 Ma) (Dietz 1961) and Sudbury (~250 km; 1,850 ± 3 Ma) (Dietz 1964). Alternatively these structures were regarded as crypto explosion features consequent on volcanic gas explosion (Nicolaysen and Ferguson 1990). However, shock metamorphic parameters indicate pressures of >10 GPa, exceeding pressures induced by volcanic explosions, nor are contemporaneous volcanic rocks associated with mega-impact structures. It has taken more than 20 years to establish the asteroid impact origin of these structures, identified by shatter cones, planar deformation features in quartz, high pressure phases [coesite, stishovite], shock vitrification [such as produce silica glass (Lechatelierite) and feldspar glass (maskelynite)], impact melting, melt breccia, pseudotachylite veins and dykes, iridium anomalies and numerous other features diagnostic of shock metamorphism).

Despite acceptance of the diagnostic hallmarks of impact outlined above, few suspected that, following the Late Heavy Bombardment of the Moon (LHB ~3.95–3.85 Ga) (Ryder 1990, 1991, 1997), extraterrestrial impacts continued to play a

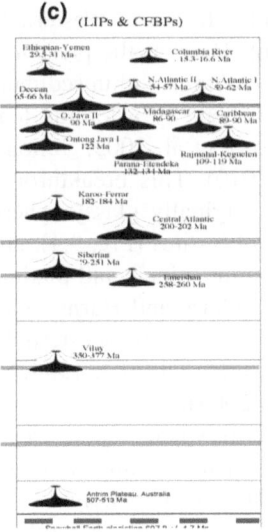

Fig. 1.2 (continued)

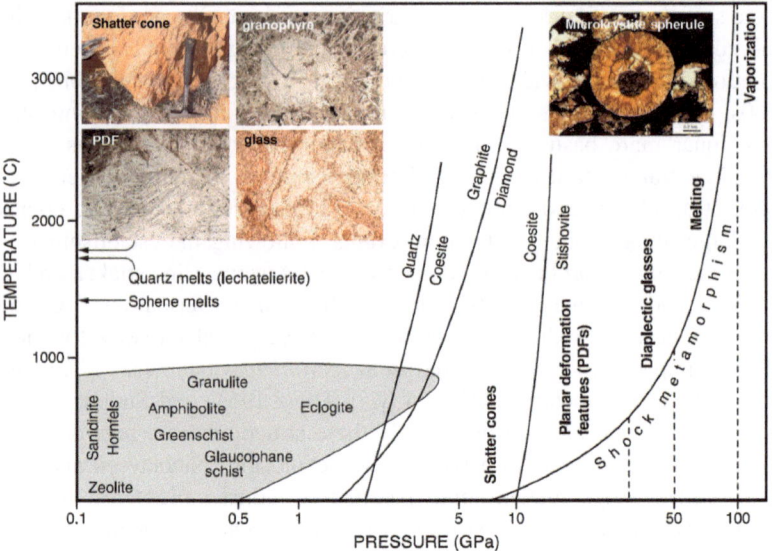

Fig. 1.3 Pressure–temperature diagram comparing conditions of shock metamorphism and conditions of endogenic crustal metamorphism. The shock-metamorphic field (from ~7 to >100 GPa) is distinct from the endogenic field is (P < 5 GPa, T < 1,000 °C). Stability curves for high-pressure minerals (coesite, diamond, stishovite) are shown for static equilibrium conditions (after French 1998, Fig. 4.1; by permission). Inset microphotographs display (1) a shatter cone; (2) granophyre core and radiating crystallites; (3) planar deformation features in quartz; (4) devitrified glass (1–3 from the Yarrabubba impact structure) and (5) a microkrystite spherule formed by condensation of impact-released vapor (Jeerinah Impact Layer)

major role in the history of Earth. However, since the 1980s geological field work in the ancient cratons of South Africa and Western Australia by Don Lowe, Gary Byerly, Bruce Simonson, Scott Hassler, the present author and other have identified major asteroid impact ejecta units within sedimentary and volcanic sequences, recording repeated impact clusters by asteroids tens of kilometers in diameter between ~3.5 and 2.5 Ga (Lowe et al. 2003; Simonson and Glass 2004; Glikson 2008; Glikson and Vickers 2010). This breakthrough was allowed by the identification of millimeter-scale originally glassy spherules in sediments of the KT impact boundary termed microkrystites (Fig. 9.4), characterized by inward radiating quench crystallites, chromium spinels and platinum group element anomalies, markedly high iridium levels (Glass and Burns 1988). Geochemical calculations and spherule size frequency analysis suggest asteroids as large as 20–50 km across (Melosh and Vickery 1991; Byerly and Lowe 1994; Shukolyukov et al. 2000; Kyte et al. 2003; Glikson and Allen 2004).

Microkrystite spherules form when vapor, ejected from craters upon large impact, condense in the atmosphere. On impact, target rocks are fragmented, shattered, fused and vaporized. The crust underlying the crater rebounds elastically, forming a dome, a process analogous to the upward ejection of a water drop when a stone is thrown into a pond. The impact vapor is dispersed with the winds, cools and condenses as myriad melt droplets which solidify as tiny glass spheres, preserved in sub-marine sediments (Fig. 1.4).

Since the 1990s no fewer than 18 microkrystite-bearing ejecta and fallout units have been detected in Archaean Pilbara and Barberton greenstone belts, overlying sediments of the Hamersley Basin and Transvaal Basin (Simonson 1992) and younger sediments. Pilbara ejecta units are dated as about ~3.47 Ga (2 units), ~2.63 Ga, ~2.57 Ga, ~2.56 Ga (2 units) and ~2.48 Ga-old, and Barberton ejecta units about 4.482, 3.472, 3.445, 3.416, 3.334, 3.256, 3.243 and 3.225 Ga (2 units) (Lowe and Byerly 2010). Several of these units represent multiple impacts (Glikson 2004a, b). A spherule unit 1.85–2.13 Ga-old is reported from Greenland (Chadwick et al. 2000) and spherule units ~1.850 Ga-old are reported from Ontario, Minnesota and Michigan (Addison et al. 2005; Jirsa et al. 2008; Cannon et al. 2010). The frequency of impact ejecta units in the Barberton greenstone belt suggests frequent intermittent bombardment of the Earth since the LHB (Lowe and Byerly 2010). Given the difficulty in identifying spherule units in the field, impact frequencies documented to date inherently represent only a minimum flux, namely the 'tip of the iceberg', yielding further support to an extension of the LHB.

The question arises, what were the effects of impacts by asteroids ~10 km and larger on the Earth's crust, its structure, tectonics, magmatic activity and

Fig. 1.4 Outcrop-scale (mesoscopic) features diagnostic of asteroid impact. **a** Shatter cone from the Gosses Bluff impact structure, central Australia, forming penetrative conical radiating horse tail-shaped striated fracture pattern (courtesy Duane Hamacher); **b** Rhombohedral fracture patterns associated with shatter cones, Yarrabubba impact structure, Western Australia; **c** Suevite—crater-fill melt breccia. Large hand specimen, about 45 cm-long, of typical fresh suevite from the Ries Crater (Germany). The specimen consists of irregular and contorted individual fragments of glass (*dark*) with roughly parallel elongation and crystalline rock fragments (*light*) in a fine clastic matrix. (From French 1998, by permission)

Table 1.1 Global impact structures (diameters > 5 km) (Mostly after the World Impact Database, including modification based on the literature and the author's observations)

Name	Locality	Diameter (km)	Age (Ma)
North America			
Ames	Oklahoma, U.S.A.	16	470 ± 30
Avak	Alaska, U.S.A.	12	3–95
Beaverhead	Montana, U.S.A.	60	~600
Carswell	Saskatchewan, Canada	39	115 ± 10
Calvin	Michigan, USA	8.5	450 ± 10
Charlevoix	Quebec, Canada	54	342 ± 15*
Chesapeake Bay	Virginia, U.S.A.	~85	35.5 ± 0.3
Chicxulub	Yucatan, Mexico	170	64.98 ± 0.05
Clearwater East	Quebec, Canada	26	290 ± 20
Clearwater West	Quebec, Canada	36	290 ± 20
Cloud Creek	Wyoming, USA	7	190 ± 30
Couture	Quebec, Canada	8	430 ± 25
Crooked Creek	Missouri, U.S.A.	7	320 ± 80
Decaturville	Missouri, U.S.A.	6	<300
Deep Bay	Saskatchewan, Canada	13	99 ± 4
Des Plaines	Illinois, U.S.A.	8	<280
Eagle Butte	Alberta, Canada	10	<65
Elbow	Saskatchewan, Canada	8	395 ± 25
Glover Bluff	Wisconsin, U.S.A.	8	<500
Gow	Saskatchewan, Canada	5	<250
Haughton	Nunavut, Canada	23	39
Kentland	Indiana, U.S.A.	13	<97
La Moinerie	Quebec, Canada	8	400 ± 50
Manicouagan	Quebec, Canada	100	214 ± 1
Manson	Iowa, U.S.A.	35	73.8 ± 0.3
Maple Creek	Saskatchewan, Canada	6	<75
Marquez	Texas, U.S.A.	12.7	58 ± 2
Middlesboro	Kentucky, U.S.A.	6	<300
Mistastin	Newfoundland/Labrador,	28	36.4 ± 4
Montagnais	Nova Scotia, Canada	45	50.50 ± 0.76
Nicholson	NWT, Canada	12.5	<400
Pilot	NWT, Canada	6	445 ± 2
Presqu'ile	Quebec, Canada	24	<500
Red Wing	North Dakota, U.S.A.	9	200 ± 25
Rock Elm	Wisconsin, U.S.A.	6	<505
Saint Martin	Manitoba, Canada	40	220 ± 32
Santa Fe	New Mexico, U.S.A.	6–13	<1,200
Serpent Mound	Ohio, U.S.A.	8	<320
Sierra Madera	Texas, U.S.A.	13	<100
Slate Islands	Ontario, Canada	30	~450
Steen River	Alberta, Canada	25	91 ± 7*
Sudbury	Ontario, Canada	250	1,850 ± 3
Upheaval Dome	Utah, U.S.A.	10	<170
Wanapitei	Ontario, Canada	7.5	37.2 ± 1.2

(continued)

Table 1.1 (continued)

Name	Locality	Diameter (km)	Age (Ma)
Wells Creek	Tennessee, U.S.A.	12	200 ± 100
Wetumpka	Alabama, U.S.A.	6.5	81.0 ± 1.5
Europe			
Boltysh	Ukraine	24	65.17 ± 0.64
Dellen	Sweden	19	89.0 ± 2.7
Gardnos	Norway	5	500 ± 10
Ilyinets	Ukraine	8.5	378 ± 5*
Kärdla	Estonia	7	~455
Keurusselkä	Finland	30	<1,800
Lappajärvi	Finland	23	73.3 ± 5.3
Lockne	Sweden	7.5	455.00
Logoisk	Belarus	15	42.3 ± 1.1
Lumparn	Finland	9	~1,000
Maniitsoq	Greenland	>100	~2,975
Mien	Sweden	9	121.0 ± 2.3
Mizarai	Lithuania	5	500 ± 20
Mjølnir	Norway	40	142.0 ± 2.6
Neugrund	Estonia	8	~470
Obolon'	Ukraine	20	169 ± 7
Paasselkä	Finland	10	<1,800
Ries	Germany	24	15.1 ± 0.1
Rochechouart	France	23	214 ± 8
Sääksjärvi	Finland	6	~560
Söderfjärden	Finland	6.6	~600
Ternovka	Ukraine	11	280 ± 10
Vepriai	Lithuania	8	>160 ± 10
Asia			
Beyenchime-Salaatin	Russia	8	40 ± 20
Bigach	Kazakhstan	8	5 ± 3
Chiyli	Kazakhstan	5.5	46 ± 7
Chukcha	Russia	6	<70
Dhala	India	11	1,700–2,100
El'gygytgyn	Russia	18	3.5 ± 0.5
Jänisjärvi	Russia	14	700 ± 5
Jebel Waqf as Suwwan	Jordan	5.5	56–37
Kaluga	Russia	15	380 ± 5
Kamensk	Russia	25	49.0 ± 0.2
Kara	Russia	65	70.3 ± 2.2
Kara-Kul	Tajikistan	52	<5
Karla	Russia	10	5 ± 1
Kursk	Russia	6	250 ± 80
Logancha	Russia	20	40 ± 20
Popigai	Russia	100	35.7 ± 0.2
Puchezh-Katunki	Russia	80	167 ± 3
Ragozinka	Russia	9	46 ± 3
Suavjärvi	Russia	16	~2,400

(continued)

Table 1.1 (continued)

Name	Locality	Diameter (km)	Age (Ma)
South America			
Araguainha	Brazil	40	244.40 ± 3.25
Serra da Cangalha	Brazil	12	<300
Vargeão Dome	Brazil	12	<70
Vista Alegre	Brazil	9.5	<65
Africa			
Aorounga	Chad	12.6	<345
Bosumtwi	Ghana	10.5	1.07
Gweni-Fada	Chad	14	<345
Luizi	DRCongo	17	<573
Morokweng	South Africa	70	145.0 ± 0.8
Oasis	Libya	18	<120
Vredefort	Orange Free State	298	2023
Australia			
Acraman	South Australia	90	~590
Amelia Creek	Northern Territory	~20	1,640–600
Connolly Basin	Western Australia	9	<60
Crawford	South Australia	8.5	>35
Flaxman	South Australia	10	>35
Foelsche	Northern Territory	6	>545
Glikson	Western Australia	~19	<508
Gnargoo (probable impact)	Western Australia	75	Post-Carboniferous
Goat Paddock	Western Australia	5.1	<50
Gosses Bluff	Northern Territory	24	142.5 ± 0.8
Kelly West	Northern Territory	10	>550
Lawn Hill	Queensland	18	>515
Matt Wilson	Northern Territory	~7.5	1,402 ± 440
Mount Ashmore	Timor Sea	>50	Eocene–Oligocene boundary
Piccaninny	Western Australia	7	<360
Shoemaker	Western Australia	30	1,630 ± 5
Spider	Western Australia	13	>570
Strangways	Northern Territory	25	646 ± 42
Talundilly	Queensland	84	128 ± 5
Tookoonooka	Queensland	55	128 ± 5
Warburton (probable impact)	South Australia	~200	~200?
Woodleigh	Western Australia	120	364 ± 8
Yarrabubba	Western Australia	>50	<2,650

Table 1.2 Precambrian impact fallout units and impact structures

Geological unit	Composition	Age	Reference
ACM-1 Mt Ada Basalt, Warrawoona Group, Pilbara Craton	Silica–sericite spherules in m-thick chert breccia/conglomerate. Overlain by felsic hypabyssal/volcanics	3,470.1 ± 1.9 Ma	Byerly et al. (2002)
ACM-2, Mt Ada Basalt, Warrawoona Group, Pilbara Craton	Silica–sericite spherules within 14 m-thick chert, arenite. Overlain by ~10 m-thick jaspilite overlying spherule unit ACM-1	3,470.1 ± 1.9 Ma	Byerly et al. (2002)
BGB-S1A and BGB-S1B, upper Hooggenoeg Formation, Onverwacht Group, Kaapvaal Craton	Two units of silica–chert spherules within 30–300 cm-thick unit of chert and arenite	3,470.4 ± 2.3 Ma	Byerly et al. (2002)
BGB-S2, base of the Mapepe Formation, Fig Tree Group, Kaapvaal Craton	310 cm-thick silica–sericite spherules. Overlain by Manzimnyama Jaspilite Member: BIF/jaspilite/ferruginous shale (520 m) and shale above BGB-S2	3,258 ± 3 Ma	Lowe et al. (2003)
BGB-S3 and BGB-4 lower Mapepe Formation, Fig Tree Group, Kaapvaal Craton	S3: 10–15 cm-thick to locally 2–3 m-thick silica–Cr-sericite–chlorite spherules, overlain by ferruginous sediments of the Ulundi Formation in the northern part of the BGB S4: 15 cm-thick arenite with chlorite-rich spherules	3,243 ± 4–3,225 ± 3 Ma	Lowe et al. (2003)
Maniitsoq impact structure, SW Greenland	A central crush zone enveloped by gneiss and cut by mafic bodies	~2,975 ± 6 Ma,	Garde et al. (2012)
JIL, top Jeerinah Formation, Fortescue Group, Hamersley Basin	Hesta: 80 cm-thick carbonate–chlorite spherules and spherule-bearing breccia; 60 cm thick overlying debris flow; overlain by Marra Mamba Iron-formation, immediately above ~60 cm-thick shale unit overlying JIL	<2,629 ± 5 Ma, >2,597 ± 5 Ma	Simonson et al. (2000a, b); Trendall et al. (2004)
Base of Carawine Dolomite, Hamersley Group, Hamersley Basin	Carbonate megabreccia-hosted microkrystite spherules; K-feldspar–carbonate–chlorite spherules in tsunami-generated carbonate–chert megabreccia	2,630 ± 6 Ma	Simonson and Hassler (1997); Rasmussen et al. (2005)

(continued)

Table 1.2 (continued)

Geological unit	Composition	Age	Reference
Monteville Formation, West Griqualand Basin, west Kaapvaal Craton	5 cm-thick spherule layer. Carbonate hosted	<2,650 ± 8 Ma, ~2,647 ± 30 Ma	(Simonson and Glass 2004); Simonson et al. (2010)
Reivilo Formation, West Griqualand Basin, western Kaapvaal Craton	1.8 cm-thick spherule unit. Carbonate-hosted	>2,581 ± 9 Ma, <2,588 ± 6	Simonson et al. (2010)
Paraburdoo Spherule Layer, Hamersley Basin, Western Australia	2 cm-thick altered spherule unit in carbonates	>2,561 ± 8 Ma, <2,597 ± 5 Ma	Hassler et al. (2011)
SMB-1, top of Bee Gorge Member, upper Wittenoom Formation, Hamersley Group, Hamersley Basin	5 cm-thick K-feldspar–carbonate–chlorite spherules in carbonate turbidite. Overlain by ferruginous siltstone (Sylvia Formation), banded iron-formation (Bruno Member)	2,541 ± 18/16 Ma	Simonson et al. (2010); Glikson (2004a, b); Trendall et al. (2004)
SMB-2, top of Bee Gorge Member, upper Wittenoom Formation, Hamersley Group, Hamersley Basin	20 cm-thick K-feldspar–carbonate–chlorite spherules within turbidite. Overlain by ferruginous siltstone (Sylvia Formation), banded iron-formation (Bruno Member)	2,541 ± 18/16 Ma	Glikson (2004a, b); Trendall et al. (2004)
S4 Shale Macroband, Dales Gorge Member, Brockman Iron-Formation Hamersley Group, Hamersley Basin	10–20 cm K-feldspar–stilpnomelane spherules at top of 2–3 m of ferruginous volcanic tuffs. Located 38 m above the base of the Brockman	2,481 ± 4 Ma	Trendall et al. (2004); Simonson et al. (2010)
Lower Kuruman Formation, West Griqualand Basin, west Kaapvaal Craton	1 cm-thick spherule unit overlain by 80 cm breccia. Located 37 m above base of banded ironstones	<2,516 ± 4 Ma	Simonson et al. (2010)
Yarrabubba, Murchison Goldfields, Western Australia	Central granophyre plug emplaced in shatter-coned PDF-bearing granite	<2.4–2.0 Ga	Macdonald et al. (2003)
Vredefort impact structure, Transvaal	Central basement plug emplaced in sedimentary–volcanic collar	2,023 ± 4 Ma	Kamo et al. (1996)
Graensco, Vallen, Ketilidean, southwest Greenland	20 cm-thick spherule unit. Carbonate-hosted spherules	~2.13–1.85 Ga	Chadwick et al. (2000)
Sudbury impact structure and related ejecta	D ~ 250 km	1,850 ± 1 Ma	Addison et al. (2005)

sedimentation? In 1989 Don Lowe and Gary Byerly observed a juxtaposition between multiple ~3.26 to 3.22 Ga-old impact ejecta units found in the Barberton greenstone belt, South Africa, and an unconformity separating an underlying thick ultramafic volcanic 'komatiite' sequences and overlying semi-continental sandstones, siliceous volcanics, banded ironstone and conglomerate (Lowe et al. 1989; Byerly et al. 1993, 1996). Ultramafic volcanics and chert underlying the impact ejecta units display fractures and dykes which signify the effects of seismic shock and tsunami waves related to the impacts (Lowe and Knauth 1977; Lowe et al. 2003). A search for equivalents ~3.26–3.22 Ga impact ejecta units in the Pilbara Craton, Western Australia, uncovered contemporaneous breaks between mafic-ultramafic volcanic sequences and semi-continental sediments, including sandstone, conglomerate and major boulder deposits, including blocks up to 250 m large (Glikson and Vickers 2006). These breaks were accompanied by intensive plutonic granitic activity in both the Pilbara and the Kaapvaal cratons.

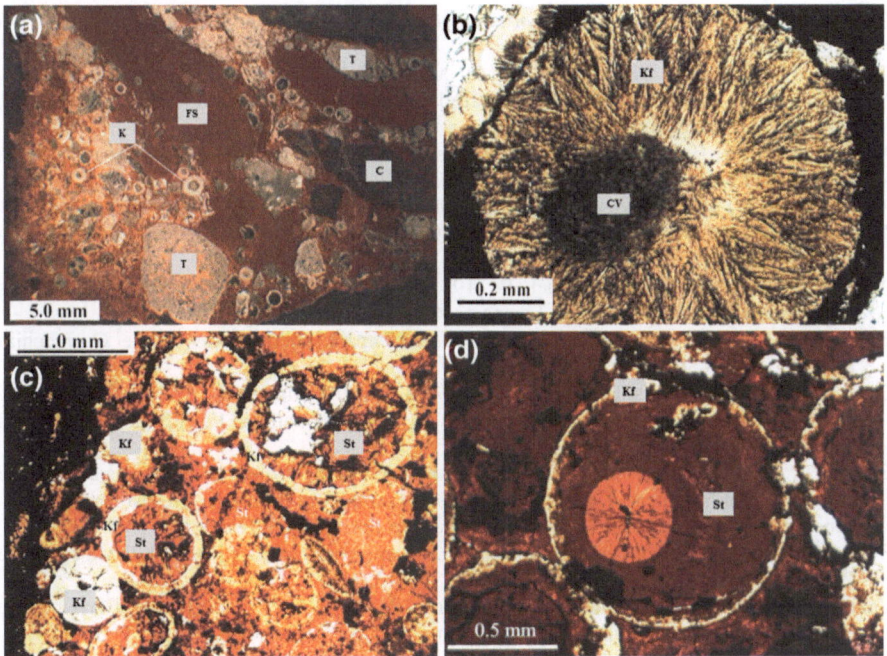

Fig. 1.5 Microkrystite spherules—devitrified and altered spherulitic glass condensates from impact-released vapor and melt (Glass and Burns 1988). **a** Microkrystite spherules and larger irregularly shaped microtektites [*T*] mixed with fragments of ferruginous shale [*FS*] and chert [*C*], base of the Jeerinah Impact Layer [*JIL*], central Pilbara Craton, Western Australia. **b** JIL Microkrystite spherule displaying inward-radiating K-feldspar crystallites [*Kf*]. The inward radiation of quench crystallites is diagnostic of impact-produced spherules, as distinct from volcanic fire-fountaining spherules which display outward-radiating crystallites. **c** Microkrystite spherules in the Dales Gorge Macroband 4 impact unit [*DGS*4], consisting of red stilpnomelane [*St*] and yellow K-feldspar shells [*Kf*]. **d** A DGS4 microkrystite spherule cored by central bubble of well crystallized stilpnomelane in matrix of cryptocrystalline stilpnomelane rimmed by K-feldspar shell

It follows that impact clusters triggered an abrupt transformation in crustal conditions, terminating the development of komatiite-rich basaltic oceanic-like crust, causing melting in the underlying rocks with consequent felsic plutonic and volcanic magmatism. Further evidence for magmatic activity triggered by impacts is observed where ~3.26–2.56 Ga-old ejecta and fallout layers are capped by iron-rich chert and banded iron formations, some of which are the source of Pilbara iron ores, hinting at the existence of large volcanic terrains from which the iron could be derived (Glikson 2006; Glikson and Vickers 2007).

To date 119 impact craters and structures larger than 5 km in diameter (Table 1.1) and 18 impact ejecta and fallout units (impactites[2]) (Table 1.2; Fig. 1.5) have been documented.

References

Abbott DH, Isley AE (2002) Extraterrestrial influences on mantle plume activity. Earth Planet Sci Lett 205:53–62
Addison WD, Brumpton GR, Vallini DA, McNaughton NJ, Davis DW, Kissin SA, Fralick PW, Hammond AL (2005) Discovery of distal ejecta from the 1850 Ma Sudbury impact event. Geology 33:193–196
Alt AD, Sears JW, Hyndman DW (1988) Terrestrial Mare: the origins of large basalt plateaus hotspot tracks and spreading ridges Journal. Geology 96:647–662
Alvarez W (1997) Tyranosaurus Rex and the crater of doom. Princeton University Press, p 185
Alvarez L, Alvarez W, Asaro F, Michel HV (1980) Extraterrestrial cause for the Cretaceous-Tertiary extinction. Science 208:1095–1108
Alvarez L, Alvarez W, Asaro F, Michel HV (1982) Iridium anomaly approximately synchronous with terminal Eocene extinctions. Science 216:886–888
Byerly GR, Lowe DR (1994) Spinels from Archaean impact spherules. Geochim et Cosmochim Acta 58:3469–3486
Byerly GR, Kröner A, Lowe DR, Walsh MM (1993) Sequential magmatic evolution of the early Archaean Onverwacht Group: evidence from the upper formations: Eos (Trans Am Geophys Union) 74:660
Byerly GR, Kröner A, Lowe DR, Todt W, Walsh MW (1996) Prolonged magmatism and time constraints for sediment deposition in the early Archaean Barberton greenstone belt: evidence from the upper Onverwacht and Fig Tree Groups. Precam Res 78:125–138
Byerly GR, Lowe DR, Wooden JL, Xie X (2002) A meteorite impact layer 3470 Ma from the Pilbara and Kaapvaal Cratons. Science 297:1325–1327
Cannon WF, Schulz KJ, Wright J, Horton D, Kring A (2010) The Sudbury impact layer in the Paleoproterozoic iron ranges of northern Michigan, USA. Geol Soc Am Bull 122:50–75
Chadwick B, Claeys P, Simonson BM (2000) New evidence for a large Palaeoproterozoic impact Spherules in a dolomite layer in the Ketilidian orogen South Greenland. J Geol Soc London 158:331–340
Chao ECT (1967) Shock effects in certain rock-forming minerals. Science 156:192–202
Chao ECT (1968) Pressure and temperature histories of impact-metamorphosed rocks based on petrographic observations. In: French BM, Short NM (eds) Shock Metamorphism of Natural Materials. Mono Book Corp Baltimore. 135-158
Darwin C (1859) On the origin of species by means of natural selection or the preservation of favored races in the struggle for life. John Murray, London, p 502

---

[2] The term impactite refers here to impact ejecta and fallout units.

Dietz RS (1961) Vredefort ring structure: meteorite impact scar? J Geol 69:496–505

Dietz RS (1964) Sudbury structure as an astroblemes. J. Geol 72:412–434

French BM (1998) Traces of catastrophe—a handbook of shock metamorphic effects in terrestrial meteorite impact structures. Lunar Planet Sci Instit Contrib 954:120

Garde AA, McDonald I, Dyck B, Keulen N (2012) Searching for giant ancient impact structures on Earth: the Meso-Archaean Maniitsoq structure, West Greenland. Earth Planet Sci Lett 2012:337–338

Glass BP, Burns CA (1988) Microkrystites: a new term for impact-produced glassy spherules containing primary crystallites Proc Lunar Planet Sci Conf XVIII: 455–458

Glikson AY (2004a) Bedout: a possible end-Permian impact crater Offshore of northwestern Australia. Science 306:613

Glikson AY (2004b) Early Precambrian asteroid impact-triggered tsunami: excavated seabed debris flows exotic boulders and turbulence features associated with 3.47–2.47 Ga-old asteroid impact fallout units, Pilbara Craton. Western Australia. Astrobiology 4:1–32

Glikson AY (2005) Geochemical and isotopic signatures of Archaean to early Proterozoic extraterrestrial impact ejecta/fallout units. Aust J Earth Sci 52:785–799

Glikson AY (2006) Asteroid impact ejecta units overlain by iron rich sediments in 3.5–2.4 Ga terrains Pilbara and Kaapvaal cratons: accidental or cause–effect relationships? Earth Planet Sci Lett 246:149–160

Glikson AY (2008) Field evidence of *Eros*-scale asteroids and impact-forcing of Precambrian geodynamic episodes Kaapvaal (South Africa) and Pilbara (Western Australia) Cratons. Earth Planet Sci Lett 267:558–570

Glikson AY (2009) Mass extinction of species: the role of external forcing. J Cosmology 2:230–234

Glikson AY, Allen C (2004) Iridium anomalies and fractionated siderophile element patterns in impact ejecta, Brockman Iron Formation, Hamersley Basin, Western Australia: evidence for a major asteroid impact in *simatic* crustal regions of the early Proterozoic earth. Earth Planet Sci Lett 20:247–264

Glikson AY, Vickers J (2006) The 3.26–3.24 Ga Barberton asteroid impact cluster: Tests of tectonic and magmatic consequences Pilbara Craton Western Australia. Earth Planet Sci Lett 241:11–20

Glikson AY, Vickers J (2007) Asteroid mega-impacts and Precambrian banded iron formations: 2.63 Ga and 2.56 Ga impact ejecta/fallout at the base of BIF/argillite units Hamersley Basin Pilbara Craton Western Australia. Earth Planet Sci Lett 254:214–226

Glikson AY, Vickers J (2010) Asteroid impact connections of crustal evolution. Aust J Earth Sci 57:79–95

Gradstein FM, Ogg JG, Smith AG, Bleeker W, Laurens LJ (2004) A new geologic timescale with special reference to Precambrian and Neogene. Episodes 72:83–100

Grey K (2005) Ediacaran Palynology of Australia. Assoc Australas Palaeontol Mem 31:439 Canberra

Grey K, Walter MR, Calver CR (2003) Neoproterozoic biotic diversification: snowball Earth or aftermath of the Acraman impact? Geology 5:459–462

Hassler SW, Simonson BM, Sumner DY, Bodin L (2011) Paraburdoo spherule layer, Hamersley Basin, Western Australia: Distal ejecta from a fourth large impact near the Archaean-Proterozoic boundary. Geology 39:307–310

Hutton J (1788) Theory of the Earth. Trans Roy Soc Edinburgh, 1(2): 209–304

Jirsa MA, Weiblen PW, Vislova T, McSwiggen PL (2008) Sudbury impactite layer near Gunflint Lake, NE Minnesota. Instit Lake Superior Geol Proc 54:42–43

Kamo SL, Reimold WU, Krogh TE, Colliston WP (1996) A 2.023 Ga age for the Vredefort impact event and a first report about shock metamorphosed zircons in pseudotachylitic breccias and granophyre. Earth Planet Sci Lett 144:369–388

Keller G (2005) Impacts volcanism and mass extinction: random coincidence or cause and effect? Aust J Earth Sci 52:725–757

Kyte FT, Shukolyukov A, Lugmair GW, Lowe DR, Byerly GR (2003) Early Archaean spherule beds: chromium isotopes confirm origin through multiple impacts of projectiles of carbonaceous chondrite type. Geology 31:283–286

Lowe DR, Byerly GR (2010) Did the LHB end not with a bang but with a whimper? 41st Lunar
    Planet Sci Conf 2563pdf
Lowe DR, Knauth LP (1977) Sedimentology of the Onverwacht group (3.4 billion years), trans-
    vaal, South Africa, and its bearing on the characteristics and evolution of the early Earth. J
    Geol 85:699–723
Lowe DR, Byerly GR, Asaro F, Kyte FJ (1989) Geological and geochemical record of 3400 mil-
    lion year old terrestrial meteorite impacts. Science 245:959–962
Lowe DR, Byerly GR, Kyte FT, Shukolyukov A, Asaro F, Krull A (2003) Spherule beds 3.47–3.24
    billion years old in the Barberton Greenstone Belt, South Africa: a record of large meteorite
    impacts and their influence on early crustal and biological evolution. Astrobiology 3:7–48
Lyell C (1830) The principles of geology, vol 2. Murray, London
Macdonald FA, Bunting JA, Cina SE (2003) Yarrabubba—a large deeply eroded impact structure
    in the Yilgarn Craton Western Australia. Earth Planet Sci Lett 213:235–247
Melosh HJ, Vickery AM (1991) Melt droplet formation in energetic impact events. Nature
    350:494–497
Nicolaysen LO, Ferguson J (1990) Cryptoexplosion structures shock deformation and siderophile
    concentration related to explosive venting of fluids associated with alkaline ultramafic mag-
    mas. Tectonophysics 171:303–335
Oberbeck VR, Marshall JR, Aggarval H (1992) Impacts tillites and the breakdown of
    Gondwanaland. J Geol 101:1–19
Rasmussen B, Blake TS, Fletcher IR (2005) U-Pb zircon age constraints on the Hamersley spherule
    beds: evidence for a single 2.63 Ga Jeerinah–Carawine impact ejecta layer. Geology 33:725–728
Roddy DJ, Shoemaker EM (1995) Meteor Crater (Barringer Meteorite Crater) Arizona: summary
    of impact conditions. Meteoritics 30:567
Ryder G (1990) Lunar samples lunar accretion and the early bombardment of the Moon. Eos
    (Trans Am Geophys Union) 71:313–322
Ryder G (1991) Accretion and bombardment in the Earth–Moon system: the lunar record. Lunar
    Planet Sci Instit Contrib 746:42–43
Ryder G (1997) Coincidence in the time of the Imbrium Basin impact and Apollo 15 Kreep vol-
    canic series: impact induced melting? Lunar Planet Sci Instit Contrib 790:61–62
Sepkoski JJ (1996) Patterns of Phanerozoic extinction: a perspective from global data bases. In:
    Walliser OH (ed) Global events and event stratigraphy. Springer-Verlag, Berlin, pp 35–52
Shoemaker EM, Kieffer SW (1979) Guidebook to the Geology of Meteor Crater Arizona. Center
    for Meteorite Studies. Arizona State University, Tempe Arizona, p 45
Shukolyukov A, Kyte FT, Lugmair GW, Lowe DR, Byerly GR (2000) The oldest impact deposits
    on Earth. In: Koeberl C, Gilmour I (eds) Lecture notes in Earth science 92: Impacts and the
    Early Earth. Springer, Berlin, pp 99–116
Simonson BM (1992) Geological evidence for an early Precambrian microtektite strewn field in
    the Hamersley Basin of Western Australia. Geol Soc Am Bull 104:829–839
Simonson BM, Glass BP (2004) Spherule layers—records of ancient impacts. Ann Rev Earth
    Planet Sci 32:329–361
Simonson BM, Hassler SW (1997) Revised correlations in the early Precambrian Hamersley
    Basin based on a horizon of resedimented impact spherules. Aust J Earth Sci 44:37–48
Simonson BM, Davies D, Hassler SW (2000a) Discovery of a layer of probable impact melt
    spherules in the late Archaean Jeerinah Formation, Fortescue Group, Western Australia. Aust
    J Earth Sci 47:315–325
Simonson BM, Hornstein M, Hassler SW (2000b) Particles in late Archean Carawine Dolomite,
    Western Australia, resemble Muong Nong-type tektites. In: Gilmour I, Koeberl C (eds)
    Impacts and the Early Earth. Springer-Verlag, Berlin, pp 181–214
Simonson BM, Hassler SW, Beukes NJ, Sumner DY (2010) Large impacts around the Archaean-
    Proterozoic boundary—an update. 41st Lunar Planet Sci Conf, 2386.pdf
Trendall AF, Compspton W, Nelson DR, deLaeter JR, Bennett VC (2004) SHRIMP zircon ages
    constraining the depositional chronology of the Hamersley group Western Australia. Aust J
    Earth Sci 51:621–644

Chapter 2
Encounters in Space

Abstract Astronomical observations of the asteroid belt, size distribution and frequency of Earth-crossing asteroids (Apollo and Aten class asteroids) are consistent with the impact history of Earth as progressively revealed by geological and geophysical studies.

Asteroids constitute stony aggregates formed by initial accretion of dust and stony particles, as well as fragmentation associated with subsequent collisions. Nearly two million bodies form the asteroid belt, located between Jupiter and Mars at 2.15–3.3 AU (255–600 million km). These bodies constitute remnants of the original formation of the solar system. Over 50 % of the asteroid belt's mass consists of the large asteroids—*Ceres* (950 km), *Vesta* (530 km) (Fig. 2.1), *Pallas* (530–565 km) and *Hygiea* (350–500 km). More than 200 asteroids measure over 100 km in diameter and 700,000 to 1,700,000 asteroids exceed 1 km in size, the rest ranging down to dust particles. Comets, rich in dust and ice, on short-lived elliptical trajectories, arrive from the Oort Cloud on the fringes of the Solar system, travelling at about half the speed of asteroids ($\sim$25–30 km s^{-1}). Whereas comets retain solid cratered surfaces they release tails of vapor representing melting of ice, as on comet Wilde-2 (Fig. 2.2). Periodic collisions between these bodies send fragments hurtling toward Jupiter which, thanks to its huge gravity pull (24.79 m/s^2, as compared with the Earth's 9.78 m/s^2), sweeps up most of these fragments, as observed by the renowned Shoemaker-Levy-9 comet impact (Fig. 2.3), thus protecting the Earth from an extraterrestrial flux which could have arrested the development of advanced life. In an uncanny juxtaposition between mythology and science, the Romans viewed the god *Jupiter* as protector of the State.

Asteroids include the main types (NASA 2012):

1. Carbonaceous (C-type) asteroids are depleted in hydrogen and helium, have chemical ratios akin to solar composition, have low albedo (0.03–0.09), include more than 75 % of known asteroids and inhabit the main belt's outer regions;
2. Siliceous (S-type) asteroids consist of metallic iron mixed with iron- and magnesium-silicates, show high albedo of 0.10–0.22, include about 17 % of known asteroids and occupy the inner asteroid belt. S-type asteroids display various degrees of melting and segregation of metal from silicate.

A. Y. Glikson, *The Asteroid Impact Connection of Planetary Evolution*,
SpringerBriefs in Earth Sciences, DOI: 10.1007/978-94-007-6328-9_2,
© The Author(s) 2013

Fig. 2.1 The Dawn spacecraft last view of asteroid 4-Vesta's North Pole, viewed from NASA's Dawn spacecraft. http://www.space.com/17461-dawn-spacecraft-leaves-giant-asteroid-vesta.html. 4-Vesta, with a diameter of 525 km, is the second largest asteroid after Ceres (952 km) (Table 2.1) and forms about 9 % of the mass of the asteroid belt. Vesta lost about 1 % of its mass in a collision, with debris from this event represented by howardite–eucrite–diogenite (HED) meteorites. The Dawn spacecraft entered an orbit around Vesta on 16 July 2011 for a one-year exploration. (NASA)

3. Metallic (M-type) asteroids consist of metallic iron, display high albedo (0.10–0.18) and inhabit the middle part of the asteroid belt.

The total mass of asteroids would form a body of ~1,500 km diameter and 16 asteroids are of diameters 240 km or larger, as in Table 2.1. Asteroid orbits are elliptical rotating in the same direction as the Earth taking over 3–6 years to complete a full course around the Sun. Near Earth asteroids (NEA) at orbits of about 1.3 AU (~195.10^6 km) include the following categories: (a) *Amor* class asteroids cross the Mars orbit, the typical one being *Eros*; (b) *Apollo* class asteroids cross Earth's orbit with a period greater than 1 year, an example being *Geographos*; (c) *Aten* class asteroids cross Earth's orbit with a period less than 1 year, an example being *Ra-Shalom*. NEA whose orbits lie between 0.983 and 1.3 astronomical units (AU) away from the Sun pose a potential impact risk for Earth. By 2012 the number of NEA detected by NASA reached 9,252 near Earth asteroids (NEA) of which ~2,250 objects are of diameters ~100–300 m, ~2,800 objects of diameters 300–1,000 m and 850 objects of diameters >1 km (NASA 2012).

An earlier explanation of the origin of the asteroid belt has been advanced in terms of the breakdown of a planet smaller than Earth. The modern view is based

Fig. 2.2 Comet Wild-2 of 5 km diameter http://ucsdnews.ucsd.edu/newsrel/science/mcstardust. asp. An image taken from the spacecraft Stardust on the 2 January, 2004. As the spacecraft flew through this material, a special collection grid filled with aerogel-b, a novel sponge-like material that's more than 99 % empty space, gently captured samples of the comet's gas and dust. The grid was stowed in a capsule which detached from the spacecraft and parachuted to Earth on 15 January, 2006. Since then, scientists around the world have been busy analyzing the samples to learn the secrets of comet formation and our solar system's history. The samples contained Glycine, an amino acid used by living organisms to make proteins, and this is the first time an amino acid has been found in a comet. http://stardust.jpl.nasa.gov/news/news115.html. (NASA)

Fig. 2.3 The collision of 21 fragments (with diameters of up to 2 km) of the Comet Shoemaker-Levy-9 with Jupiter on 16 July, 1994, projected in advance by David Levy, Caroline Shoemaker and Eugene Shoemaker, constituted the first directly observed extraterrestrial collision in the solar system. It was photographed with a 40 cm Schmidt telescope at the Palomar Observatory in California. The fragmented nature of the comet is attributed to a previous approach to Jupiter in July 1992 within the Roche fragmentation limit. Impact speed is measured as 60 km/s http://www2.jpl.nasa. gov/sl9/. (NASA)

Table 2.1 Approximate number of asteroids N close to diameter D

D	N	Examples
100 m	~25,000,000	2011 AG5 (D ~ 140 m)
300 m	4,000,000	Apophis (D ~ 270 m)
500 m	2,000,000	2005 YU55 (D ~ 400 m)
1 km	750,000	1998 SF36 (D ~ 725 m)
3 km	200,000	Toutatis (D ~ 2.3–5.3 km)
5 km	90,000	Otawara (D ~ 5.5 km)
10 km	10,000	Gaspra (D ~ 19 × 12 × 11)
30 km	1,100	Eros (D ~ 34 × 11 × 11 km)
50 km	600	Ida (D ~ 58 × 23 km)
100 km	200	Lutetia (D ~ 100 km)
200 km	30	Iris (D ~ 213 km)
300 km	5	Interamnia (D ~ 326 km)
500 km	3	Vesta (D ~ 525 km); Pallas (D ~ 545 km); Hygiea (D ~ 431 km)
900 km	1	Ceres (D ~ 952 km)

on the compositional diversity of asteroids, suggesting the asteroid belt formed during the first 10 million years of solar history through gravitational accretion from the primitive solar disc, followed by aggregation, limited melting and fractionation. After their formation, the asteroids underwent an early stage of internal heating and surface melting due to impacts and surface erosion by the solar wind and micrometeorites. The NEAR Shoemaker mission, aimed at observing *Eros* (preface photograph), its surface features, chemistry, mineral composition, spin state, magnetic field and the effects of the solar wind. The probe carried 56 kg of equipment, including an X-ray/gamma ray spectrometer, a near-infrared spectrograph, a multispectral camera, a laser range finder and a magnetometer. *Eros* is an S-type asteroid, namely a body dominated by silicate minerals. NEAR Shoemaker's other task was to draw comparisons between the composition of this asteroid and S-Type meteorites found on Earth. The mission ended with a touchdown in the "saddle" region of *Eros*. Although NEAR Shoemaker was not designed as a Lander, the craft's gamma-ray spectrometer continued to collect data for 2 weeks on the elemental composition of *Eros*. The spacecraft made its last call to Earth on 28 February 2001.

Several space encounters with asteroids and comets followed. The Japanese *Hayabusa* probe collected asteroid debris from the asteroid *Itokawa* (640 × 270 m) on 12 September 2005. NASA's Deep Impact probe encountered the cratered nucleus of comet *Tempel-1* (100–200 m.) on 4 July 2005. On 10 July this year, the European Space Agency's probe *Rosetta* passed 3,162 km from the asteroid *Lutetia* (long axis 134 km) at a speed of 15 km/s. *Rosetta's OSIRIS* images of *Lutetia*, taken with both wide-angle and narrow-angle cameras, with resolution to 60 m, show large, eroded craters imprinted by young, well-defined craters. *Lutetia* shows spectral characteristics intermediate between those of C-Type asteroids and M-Type asteroids.

In the wake of the Late Heavy Bombardment during ~3.95–3.85 Ga impacts by very large asteroids and comets in the inner solar system declined from

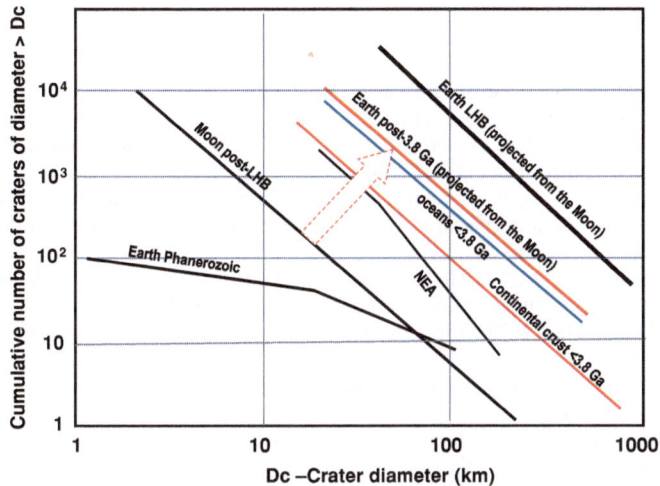

Fig. 2.4 Crater diameter versus cumulative number of craters of diameters <Dc for (1) Late heavy bombardment of Earth, extrapolated from lunar data of Barlow (1990); (2) Lunar post-Late Heavy Bombardment (*LHB*); (3) Earth post 3.8 Ga (projected from the Moon); (4) Earth oceans (~80 % of the Earth's surface); Earth continental crust (~20 % of the Earth surface); (5) Phanerozoic impact rates displaying erosional loss of small craters (modified after Grieve and Dence 1979)

a flux of 4–$9 \cdot 10^{-13}$ km^2 year^{-1} (for craters Dc $\geq$ 18 km) to a flux of 3.8–$6.3 \cdot 10^{-15}$ km^2 year^{-1} (for craters Dc $\geq$ 20 km) (Grieve and Dence 1979; Baldwin 1985; Ryder 1990). Post-LHB impact rates on Earth estimated from lunar crater counts on Mare surfaces are of the same order of magnitude as cratering rate of $5.9 \pm 3.5 . 10^{-15}$ km^{-2} year^{-1} (for craters Dc $\geq$ 20 km), as estimated from astronomical observations of near-Earth asteroids (NEA) and comets (Shoemaker and Shoemaker 1996). These authors estimated an impact flux where N ∞ Dc$^{-1.8}$ to N ∞ Dc$^{-2.0}$ (N = cumulative number of craters with diameters larger than Dc; Dc = crater diameter). Plots of crater size versus cumulative crater size frequency as projected from the Moon allow estimates of impact incidence including >100 craters of Dc $\geq$ 300 km and >50 craters of Dc $\geq$ 500 km since the LHB (Fig. 2.4).

To date a minimum of 20 impacts by asteroids >10 km in diameter has been recorded from microkrystite spherule units [~3,482, 3,472–3,467 (2), 3,445, 3,416, 3,334, 3,256, 3,243, 3,225 (2), 2.63, 2.57, 2.56 (2), 2.54, 2.48 Ga] and from large impact structures (2.975 Ga (Maniitsoq—Garde et al. 2012), 2.023 (Vredefort—Gibson and Reimold 2001), 1.85 Ga (Sudbury—Therriault et al. 2002) (Tables 1.1 and 1.2). This implies that about one third of the large asteroid influx estimated from the lunar flux (>100 craters of Dc $\geq$ 300 km, Fig. 2.4) has been identified on Earth to date (Chap. 9). These estimates do not take into account probable and possible as yet unproven impact structures such as the Warburton shock metamorphic terrain (see Sect. 11.6). Such a surprising high rate of preservation is explained by the crustal depth of major impact craters and rebound domes and the relatively high thickness of their ejecta units.

References

Baldwin RB (1985) Relative and absolute ages of individual craters and the rates of infalls on the Moon in the post-imbrium period. Icarus 61:63–91

Barlow NG (1990) Estimating the terrestrial crater production rate during the late heavy bombardment period. Lunar Planet Instit Contrib 746:4–7

Garde AA, McDonald I, Dyck B, Keulen N (2012) Searching for giant ancient impact structures on Earth: the Meso-Archaean Maniitsoq structure, West Greenland. Earth Planet Sci Lett 2012:337–338

Gibson RL, Reimold WU (2001) The vredefort impact structure South Africa: the scientific evidence and a two-day excursion guide. Council Geosci Mem 92:111

Grieve RAF, Dence MR (1979) The terrestrial cratering record: II the crater production rate. Icarus 38:230–242

NASA (2012) Near Earth Objects Program. http://neo.jpl.nasa.gov/stats/, http://nssdc.gsfc.nasa.gov/planetary/text/asteroids.txt

Ryder G (1990) Lunar samples lunar accretion and the early bombardment of the Moon. Eos (Trans Am Geophys Union) 71:313–322

Shoemaker EM, Shoemaker CS (1996) The proterozoic impact record of Australia. Aust Geol Surv Org J Aust Geol Geophys 16:379–398

Therriault AM, Anthony D, Flower R, Grieve RAF (2002) The sudbury igneous complex: a differentiated impact melt sheet. Econ Geol 97:1521–1540

Chapter 3
Lunar Impacts and the Late Heavy Bombardment in the Earth–Moon System

Abstract Growing evidence indicates the ~3.95–3.85 Ga-old Late Heavy Bombardment in the Earth-Moon system was succeeded by intermittent large impacts by asteroids of >10 km-diameter, represented by well-preserved multiple impact ejecta units in the oldest well-preserved supracrustal greenstone sequences in South Africa and Western Australia.

The period termed the *Late Heavy Bombardment* (LHB), broadly defined at 3.95–3.85 Ga, has been interpreted as representing the tail-end of planetary accretion or, alternatively, a temporally distinct bombardment episode (Ryder 1990, 1991). Early views of the LHB, marked by impact incidence of $4–9 \cdot 10^{-13}$ km^{-2} $year^{-1}$ on the Moon (for craters >18 km) (Baldwin 1985), have been questioned by Ryder (1997) in view of a scarcity of pre-LHB ejecta in the lunar highlands, which may or may not reflect the severe sampling limitations.

Some of the largest lunar Mare basins contain low-titanium basalts, which likely represent impact-triggered volcanic activity (Ryder 1997). These include the Mare Imbrium (3.86 Ga) and 3.85 ± 0.03 Ga-old K, REE, and P-rich-basalts (KREEP) (BVTP 1981). Similar relationships between impact and volcanic activity may pertain in Oceanus Procellarum (3.29–3.08 Ga basalts) and in Hadley Apennines (3.37–3.21 Ga basalts) (BVTP 1981). The likelihood of impact-volcanic relationships on the Moon gains support from the recent laser 40Ar/39Ar analyses of lunar impact spherules (Apollo 14, Fra Mauro Formation) (Muller 1993; Culler et al. 2000). Impact events tentatively indicated by the lunar spherule data include ~3.87, ~3.83, ~3.66, ~3.53 and ~3.47 Ga peaks, with large analytical error margins. A significant age spike is indicated at ~3.18 Ga, i.e. near the boundary between the Late Imbrian lunar era (~3.9–3.2 Ga) and the post-mare Eratosthenian lunar era (~3.2–1.2 Ga), as defined by the cratering record (Wilhelm 1987). Some 34 lunar impact spherules yield a mean age of 3,188 ± 198 Ma with a median age at 3,181 Ma, whereas 7 spherule ages with error <100 m.y. yield a mean age of 3,178 ± 80 Ma and a median at 3,186 Ma, hinting at a broad overlap with the 3.26–3.22 Ga impact cluster in the Barberton Mountain Land, Transvaal (Sect. 8.1.2).

Prior to the *Apollo* missions (1963–1972) the lunar craters were interpreted by some as volcanic in origin, an idea laid to rest by the impact-induced shock-metamorphic state of the lunar samples. The record of asteroid impacts

A. Y. Glikson, *The Asteroid Impact Connection of Planetary Evolution*,
SpringerBriefs in Earth Sciences, DOI: 10.1007/978-94-007-6328-9_3,
© The Author(s) 2013

observed on the Moon includes old eroded craters such as the *South Pole Aitken* (2,240 km; 13 km-deep). The period defined as the *Late Heavy Bombardment* (LHB) includes 3.95–3.85 billion years-old [Ga] craters flooded by basaltic lava, termed Mare, representing a distinct episode of bombardment by asteroids many tens of kilometers to near one hundred kilometers in diameter. Principal craters formed during the LHB include *Imbrium* (outer ring—1,300 km), *Orientale* (930 km), *Tranquillitatis* (873 km), *Serenitatis* (707 km), *Crisium* (605 km) and *Smythii* (373 km). The Lunar impact history is classified in terms of several stages (Wilhelm 1987):

a *pre-Nectarian era* (>3.9 Ga).
b *Nectarian era* (3.92–3.85 Ga, named after the *Nectaris* impact basin—333 km).
c *Imbrian era* (3.85–3.2 Ga).
d *Eratosthenian era* (3.2–1.1 Ga).
e *Copernican era* (1.1–0 Ga).

Craters of the two latter periods, imprinted on the flat surfaces of the large lunar mare, display radial ejecta rays, i.e. *Tycon, Copernicus*.

The growing evidence for multiple impact ejecta units in the oldest well preserved supracrustal greenstone sequences militates for ongoing heavy bombardment (Lowe and Byerly 2010). On the other hand no definitive evidence of the effects of 3.95–3.85 Ga impacts on the Earth have been identified to date, despite extensive records of ~3.8–4.4 Ga-old zircons in Archaean gneisses and derived clastic sediments (Wilde et al. 2001). No shock metamorphic planar deformation features (PDF) were observed to date in zircons of these early gneisses, nor have significant siderophile element anomalies been detected, with the exception of Tungsten isotope anomalies in Greenland gneisses (Schoenberg et al. 2002). Whereas coupling of the Earth-Moon system about ~4.5 Ga remains beyond reasonable doubt (Ringwood 1986), likely evidence, such as recrystallization of PDF, has been obliterated by high grade metamorphism.

In the wake of accretion of chondritic fragments and cosmic dust, gravitational collapse, melting and contemporaneous cometary contribution (Chyba 1993; Chyba and Sagan 1996; Delsemme 2000), the emergence of granitic crustal tracts is indicated by <4.4 Ga zircons (Mojzsis and Harrison 2001, Mojzsis and Harrison 2002). This signifies an onset of two-stage mantle melting processes, which require hydrous melting of mafic–ultramafic crust (Ringwood and Green 1966). From $^{18}O/^{16}O$ evidence contained in the zircons about ~4.4 Ga-old Earth may have been cool enough to allow presence of liquid water at the surface (Wilde et al. 2001; Peck et al. 2001; Mojzsis et al. 2001). Whether life existed at that stage is unknown. Earliest replicating cells probably required only twenty or so elements (da Silva and Williams 1991) at submarine hot springs, and as many fundamental organic molecular components (Wald 1964).

During a Late Heavy Bombardment of Earth (~3.95–3.85 Ga), exposure to cosmic radiation, electric and thermal flash associated with large asteroid impacts, clouding effects and acid rain would have annihilated any photosynthesizing bacterial colonies at the surface (Zahnle and Greenspoon 1990; Zahnle and Sleep

1997; Chyba 1993; Chyba and Sagan 1996), possibly with the exception of extremophile chemotrophic bacteria residing around '*black chimneys*' and '*nanobes*' (Uwins et al. 1998) in faults and fractures. Traces of $^{13}C/^{12}C$ isotopically light graphite within apatite in 3.85 Ga banded iron formation in southwestern Greenland

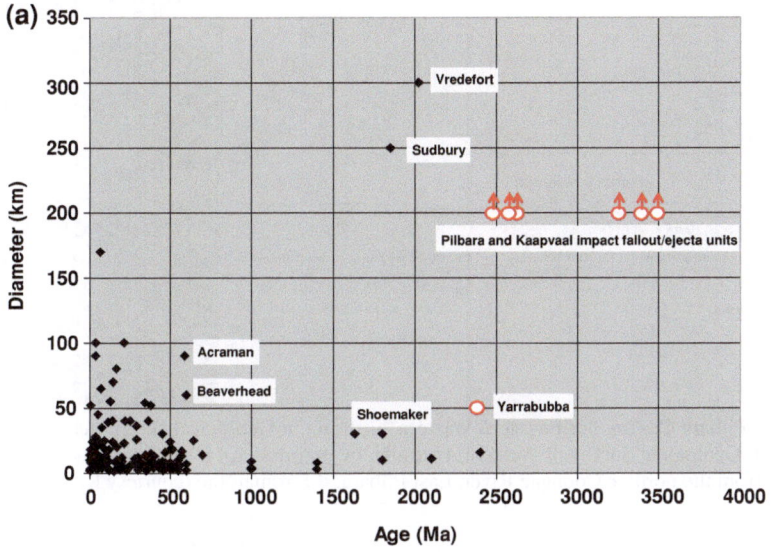

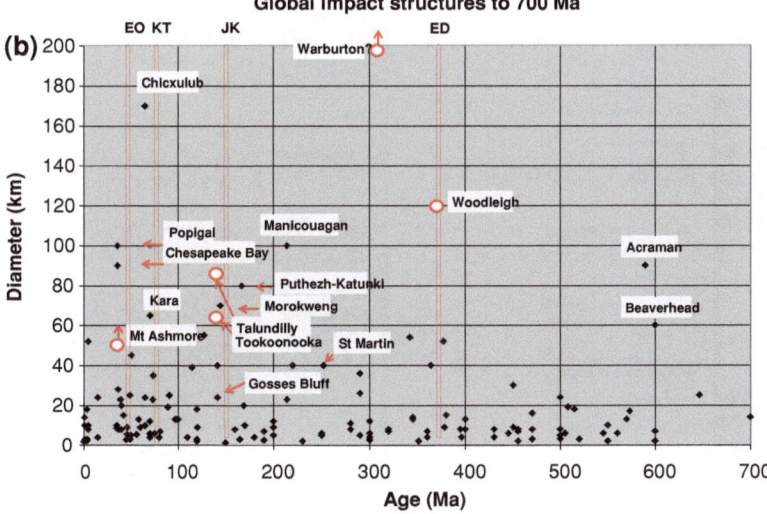

Fig. 3.1 a—Age versus diameter of global impact structures (Earth Impact Database, EID 2011) (Table 1.1) and impact ejecta units (Table 1.2). **b**—Age versus diameter of global impact structures younger than 700 Ma, based on Table 1.1 (Earth impact database, EID 2011) and highlighting large Australian impact structures

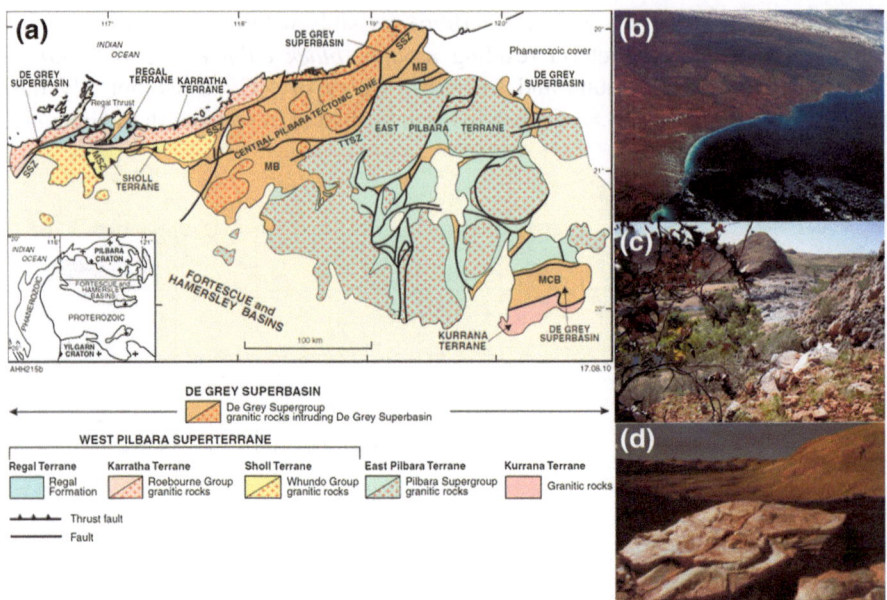

Fig. 3.2 Pilbara Craton, northwestern Western Australia. **a** Geological sketch map (courtesy A. Hickman, Geological Survey of Western Australia, by permission). **b** satellite view of the Pilbara looking from the north. **c** Coongan River, East Pilbara. **d** Stromatolite (courtesy R. Morrison)

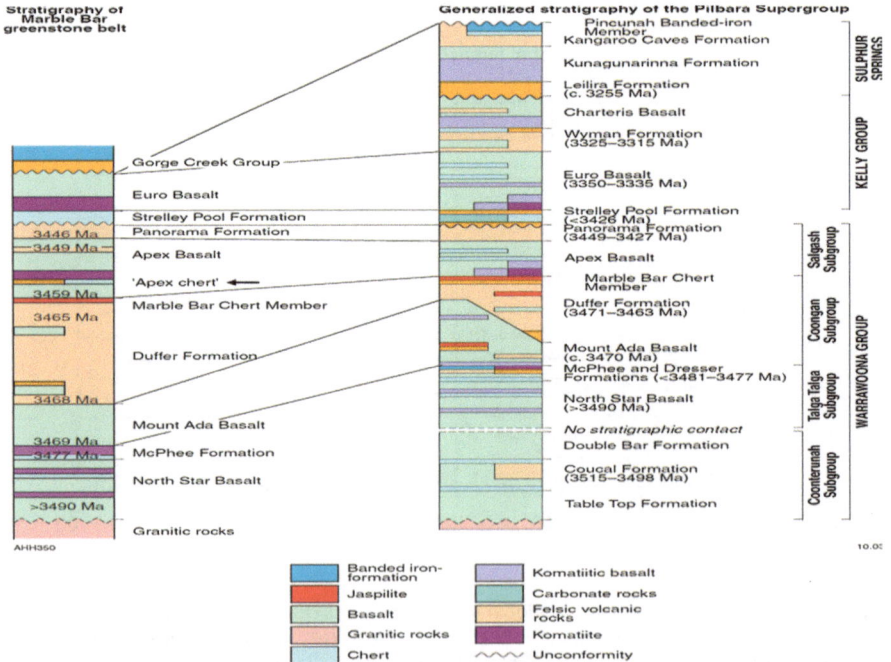

Fig. 3.3 Stratigraphy of the Marble Bar greenstone belt (Courtesy Arthur Hickman, Geological Survey of Western Australia, by permission)

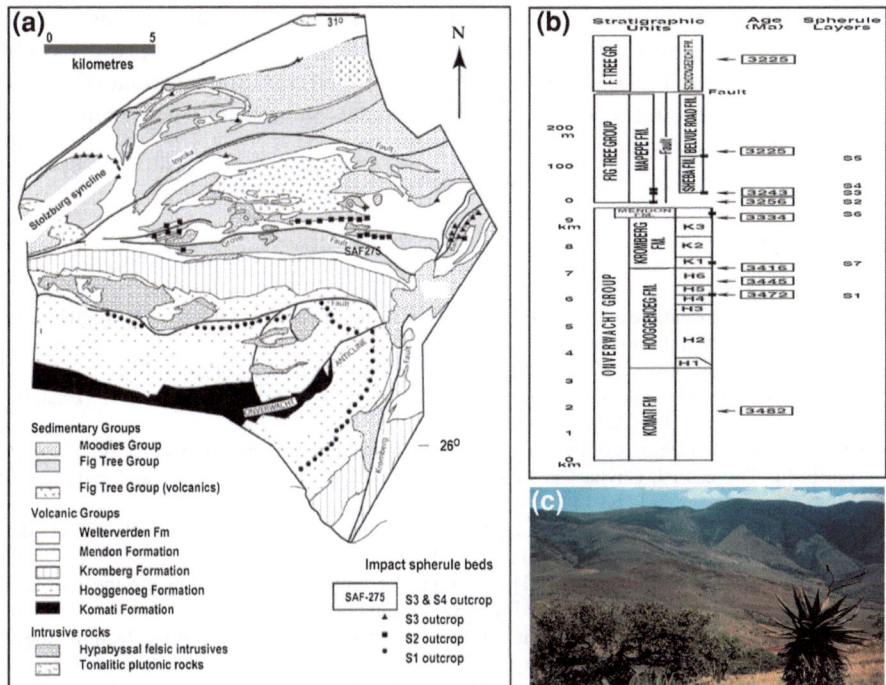

Fig. 3.4 Barberton greenstone belt (BGB), Kaapvaal Craton, South Africa. Elsevier, by permission; **a** Geological sketch map of the southern part of the BGB (Glikson 2008, after Lowe et al. 2003; Elsevier, by permission), **b** a schematic stratigraphic columnar diagram of the BGB showing the ages of impactite units (A and B from Lowe and Byerly, 2003 and 2010, by permission, and Elsevier by permission), **c** A view of the Barberton Mountain Land, looking eastward from the Kromberg syncline toward Swaziland

were suggested as possible clues for such habitat (Mojzsis et al. 1996; Mojzsis and Harrison, 2000). However, according to other workers the graphite, dispersed through the rock represents later contamination (Nutman and Friend 2006).

Continuous accretion of cosmic dust from about 3.8 Ga, estimated from deep-ocean pelagic sediment cores to have occurred at the rate of 60,000 tons/year (Kyte 2002), contributed a small fraction of about $0.2 \cdot 10^{-7}$ of the Earth mass. The post-LHB asteroids and comet flux can be roughly estimated from the impact frequency/size distribution. Based on the dominantly mafic/ultramafic composition and the scarcity of shocked quartz grains in Archaean and early Proterozoic impact ejecta, the majority of these impacts must have occurred in basaltic oceanic regions of the Earth (Simonson et al. 1998; Glikson and Allen 2004). This conclusion is in agreement with estimates of the growth of continental crust with time based on rare earth elements (Taylor and McLennan 1981) and Sm–Nd isotopes (McCulloch and Bennett 1994). As suggested by the age distribution of impact structures, ejecta and fallout units, impacts did not occur as a continuous or random flux but display a tendency to occur as clusters (Table 1.2; Fig. 3.1).

Metamorphism and anatexis in high-grade terrains generally obscures Early Precambrian impactites, with the exception of two remarkably well preserved terrains, including the Barberton Greenstone Belt (BGB) in South Africa (Fig. 3.4) and the Pilbara Craton (Figs. 3.2, 3.3), Western Australia, from where many of the examples will be derived.

References

Baldwin RB (1985) Relative and absolute ages of individual craters and the rates of infalls on the Moon in the post-imbrium period. Icarus 61:63–91

BTVP—Basaltic Volcanism Project (1981). Pergamon Press Inc, New York, p 1286

Chyba CF (1993) The violent environment of the origin of life: progress and uncertainties Geochim et Cosmochim Acta 57:3351–3358

Chyba CF, Sagan C (1996) Comets as the source of prebiotic organic molecules for the early Earth. In: Thomas PJ, Chyba CF, McKay CP (eds) Comets and the origin and evolution of life. Springer, New York, pp 147–174

Culler TS, Becker TA, Muller RA, Renne PR (2000) Lunar impact history from 39Ar/40Ar dating of glass spherules. Science 287:1785–1789

Da Silva JJRF, Williams RJP (1991) The biological chemistry of the elements: the inorganic chemistry of life. Oxford University Press, Oxford, p 600

Delsemme AH (2000) Cometary origin of the biosphere: 1999 Kuiper prize lecture. Icarus 146:313–325

Glikson AY, Allen C (2004) Iridium anomalies and fractionated siderophile element patterns in impact ejecta, Brockman Iron Formation, Hamersley Basin, Western Australia: evidence for a major asteroid impact in *simatic* crustal regions of the early Proterozoic earth. Earth Planet Sci Lett 20:247–264

Kyte FT (2002) Tracers of extraterrestrial components in sediments and inferences for Earth's accretion history. Geol Soc Am Sp Pap 356:21–38

Lowe DR, Byerly GR (2010) Did the LHB end not with a bang but with a whimper? 41st Lunar Planet Sci Conf 2563 pdf

McCulloch MT, Bennett VC (1994) Progressive growth of the Earth's continental crust and depleted mantle: geochemical constraints. Geochim et Cosmochim Acta 58:4717–4738

Mojzsis SJ, Harrison TM (2000) Vestiges of a beginning: clues to the emergent biosphere recorded in the oldest known sedimentary rocks. GSA Today 10:1–6

Mojzsis SJ, Harrison TM (2002) Establishment of a 3.83-Ga magmatic age for the Akilia tonalite southern West Greenland. Earth Planet Sci Lett 202:563–576

Mojzsis SJ, Harrison MT, Pidgeon RT (2001) Oxygen isotope evidence from ancient zircons for liquid water at the Earth's surface 4300 Myr ago. Nature 409:178–180

Muller RA (1993) Technical report LBL-34168. Lawrence Berkeley National Laboratory Berkeley CA

Nutman AP, Friend CRL (2006) Re-evaluation of oldest life evidence: infrared absorbance spectroscopy and petrography of apatites in ancient metasediments, Akilia. W Greenland Precamb Res 147:100–106

Peck WH, Valley JW, Wilde SA, Graham CM (2001) Oxygen isotope ratios and rare earth elements in 3.3 to 4.4 Ga zircons: Ion microprobe evidence for high d18O continental crust and oceans in the Early Archean. Geochim et Cosmochim Acta 65:4215–4229

Ringwood AE (1986) Origin of the Earth and Moon. Nature 322:323–328

Ringwood AE, Green DH (1966) An experimental investigation of the gabbro-eclogite transformation and some geophysical implications. Tectonophysics 3:383–427

Ryder G (1990) Lunar samples lunar accretion and the early bombardment of the Moon. Eos (Trans Am Geophys Union) 71:313–322

Ryder G (1991) Accretion and bombardment in the Earth–Moon system: the Lunar record. Lunar Planet Sci Instit Contrib 746:42–43

Ryder G (1997) Coincidence in the time of the imbrium basin impact and Apollo 15 Kreep volcanic series: impact induced melting? Lunar Planet Sci Instit Contrib 790:61–62

Schoenberg R, Kamber B, Collerson KD, Moorbath (2002) Tungsten isotope evidence from ~3.8 Gyr metamorphosed sediments for early meteorite bombardment of the Earth. Nature 418:403–405

Simonson BM, Davies D, Wallace M, Reeves S, Hassler SW (1998) Iridium anomaly but no shocked quartz from Late Archaean microkrystite layer: oceanic impact ejecta? Geology 26:195–198

Taylor SR, McLennan SM (1983) Geochemistry of Early Proterozoic sedimentary rocks and the Archaean—Proterozoic boundary. Geol Soc Am Mem 161:119–131

Uwins PJR (1998) Novel nano-organisms from Australian sandstones. Am Mineral 83:1541–1550

Wald G (1964) The origin of life. Proc Natl Acad Sci USA 52:595–611

Wilde SA, Valley JW, Peck WH, Graham CM (2001) Evidence from detrital zircons for the existence of continental crust and oceans on the Earth 4.4 Gyr ago. Nature 409:175–178

Wilhelms DE (1987) The geological history of the Moon. US Geol Surv Prof Pap 1348 p 302

Zahnle K, Grinspoon D (1990) Comet dust as a source of amino acids at the Cretaceous/Tertiary boundary. Nature 348:157–160

Zahnle K, Sleep NH (1997) Impacts and the early evolution of life In: Comets and the Origin and Evolution of Life. In: Thomas P, Chyba C, McKay C (eds) Comets and the origin of life, pp 175–208. Springer

Chapter 4
Impact and Cratering Dynamics

Abstract The classic studies of the Barringer Crater, Arizona, Ries Crater, South Germany, and more recently Chicxulub impact structure, Yucatan, and other craters, coupled with laboratory high-velocity impact experiments, have established the dynamics of impact cratering, shock metamorphism and ballistic ejecta processes.

Detailed studies of impact structures (Shoemaker 1963; Dence et al. 1977; Grieve et al. 1981; Stöffler et al. 1988) and experimental and theoretical studies (Melosh 1989; Holsapple and Schmidt 1982, 1987) resolve the essential mechanism of hypervelocity impact cratering and dispersal of ejecta. At high velocities (>11 km/second of stony meteorites >50 m-diameter) or iron meteorites (>20 m-diameter) an extraterrestrial impact results in craters 10–20 times the projectile diameter whereas smaller bodies retarded by the atmosphere result in craters with diameter somewhat larger than the projectile (French 1998). Projectiles penetrate solid rock by about 1–2 times their diameter. Impact-generated shock waves reaching several hundred GPa (Giga-Pascal) at wave velocities in excess of 10 km/second, higher than those produced by endogenic generated seismic events of <1 GPa (Melosh 1989). At the point of impact peak pressures may exceed 100 GPa and are attenuated to 10–50 GPa over kilometer-scale distances. Pressures decrease to 1–2 GPa at the margin of impact structures (Kieffer and Simonds 1980) where seismic waves decline to 5–8 km/second, a velocity analogous to that of endogenic fault-generated earthquakes and volcanic-generated waves.

The relations between the size of asteroids, the monitored number of asteroids of the particular size, the frequency of impacts and the energy released by impact are portrayed in Fig. 4.1 (after Morrison 2006). For small objects such as the ~50 m-large Tunguska comet, the number of NEA is estimated as 10^6, impact incidence 1 in 1,000 years and energy release ~10 Megaton TNT; For medium size impacts such as have resulted in the ~24 km-large Gosses Bluff, Northern Territory, and 11 km-large Spider impact structure, Western Australia, the number of NEA is estimated in the range of ~200–1,000 objects, impact incidence of one in 10^6 to 10^7 years and energy release 10^5–10^6 Megaton TNT, whereas for a KT size impact forming a ~170 km-diameter structure the number of NEA is estimated as ~10, impact frequency 1 in 100 million years and energy release ~10^8 Megaton TNT (Fig. 4.1).

A. Y. Glikson, *The Asteroid Impact Connection of Planetary Evolution*,
SpringerBriefs in Earth Sciences, DOI: 10.1007/978-94-007-6328-9_4,

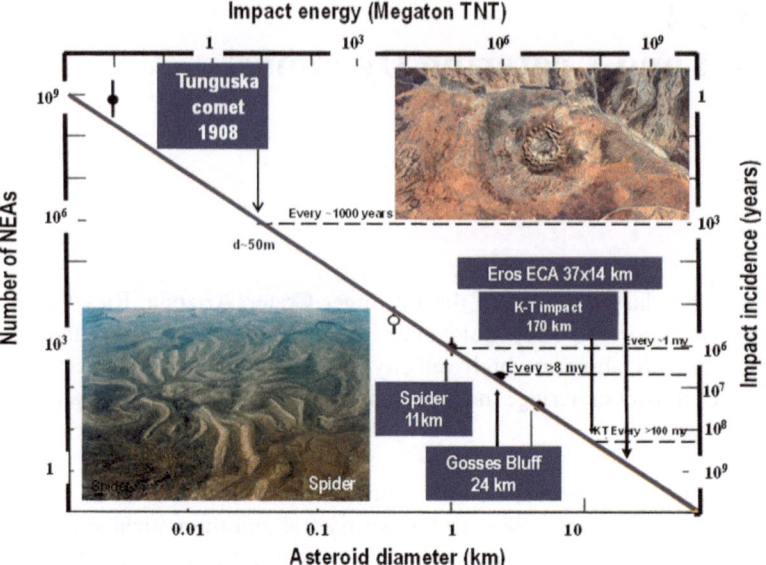

Fig. 4.1 Power-law fit to the average impact frequency for the whole Earth as a function of impact energy in megatons of TNT (after Morrison 2006, NASA). Examples of impacts on different scales are shown for the Tunguska comet (Siberia), Spider impact structure (Kimberley, northern Australia) (World Impact Database), Gosses Bluff impact structure (central Australia), the KT boundary impact and for potential impact by the asteroid *Eros*. Gosses Bluff space image (NASA); Spider photograph (Earth Impact Database, by permission)

Impact cratering is modeled in terms of three stages: compression, excavation and modification (Gault et al. 1968; Melosh 1989) (Fig. 4.2). Experimental studies of the transfer of kinetic energy to seismic shock (O'Keefe and Ahrens 1977, 1993) resolve compressive shock wave affecting the target rocks as well as rarefaction behind the projectile leading to release of pressure, melting and vaporization of parts of the projectile and interfaced rocks. Peak shock and rarefaction pressure release, melting and vaporization occur on a time scale of few seconds. The rarefaction pressure release wave extends into the surrounding rocks, with consequent fragmentation, excavation, cratering and ejection. Seismic shock is reduced exponentially with the distance, producing faulting and brecciation analogous to those associated with endogenic tectonic events. A transient bowl-shaped symmetric crater with an approximate depth/diameter ratio of 1/3 forms due to upward and inward dislocation in the upper largely ejected zones, and downward the outward dislocation in the lower zone, of the impacted zone (Grieve at al. 1977; Grieve and Cintala 1992) (Fig. 4.2). The process as modeled by Melosh (1989) suggests excavation of the ~1 km-diameter Barringer Meteor Crater in about ~6 s while that of a 200 km-diameter crater in about 90 s.

For impact craters larger than 2 km-diameter (in sediments), or 4 km (in crystalline rocks), elastic rarefaction and centripetal rebound of impact-compressed

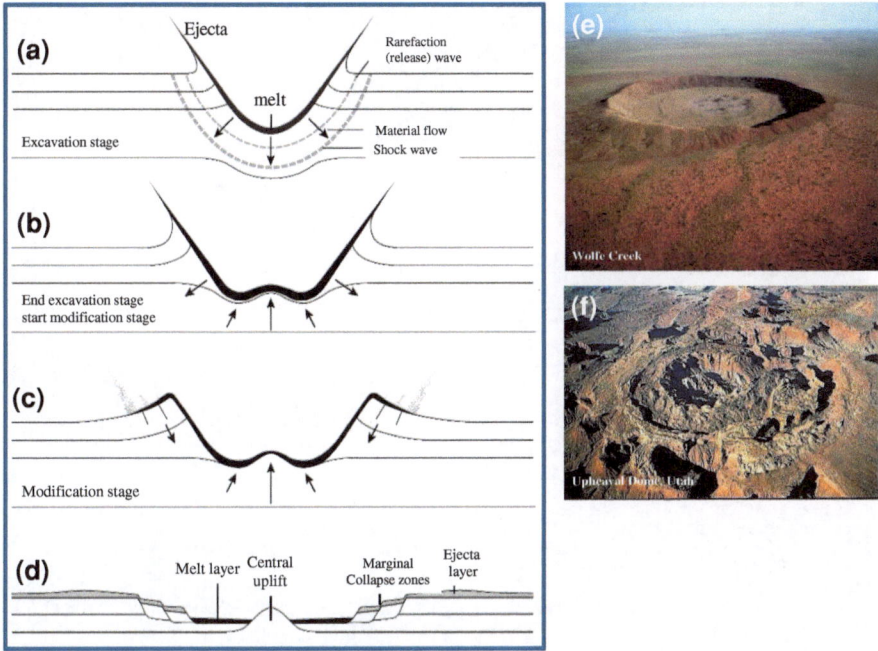

Fig. 4.2 Development of a complex impact structure (French 1998; by permission). **a** Formation of a large transient crater. **b** Initial development of central uplift during the subsequent modification stage. **c** Peripheral collapse accompanied by continuing development of the central uplift, draping of the melt layer (*black*) over the uplifted rocks. **d** Final structure consists of a central uplift surrounded by a relatively flat plain and by a terraced rim produced by inward movement along stepped normal faults. An ejecta layer (stippled) covers the target rocks around the structure. The diameter of the final structure, measured at the outer rim beyond the outermost fault, may be ×1.5–2.0 the diameter of the original transient crater. This morphology pertains to terrestrial impact structures 2–25 km in diameter; larger structures tend to develop one or more concentric rings (Fig. 4.3). **e** Simple crater: Wolfe Creek, Kimberley (R. Morrison, by permission); **f** Upheaval Dome, Utah (Jim Wark, by permission)

rocks result in formation of a central uplift dome (Dence 1968; Grieve et al. 1981) (Fig. 4.2). The process depends on the rheology of the target rocks, the shock wave characteristics and on gravity. Uplift of a central dome is accompanied by collapse along outer rim zones through concentric normal faults. Grieve and Pilkington (1996) formulated the morphometric relations between the central structural uplift (SU) and the final diameter of the structure (D) in the terms $SU = 0.086D^{1.03}$, implying an uplift of approximately >10 km for an impact structure 100–200 km in diameter, a value which depends on the rheology of the impacted rocks. An impact affecting crystalline basement rocks results in uplift of a basement plug surrounded by domed layered supracrustal rocks, for example in the Vredefort (South Africa) (Sect. 9.2) and Woodleigh (Carnarvon Basin, Western Australia) (Sect. 11.5) impact structures. An impact exclusively affecting layered supracrustal rocks produces

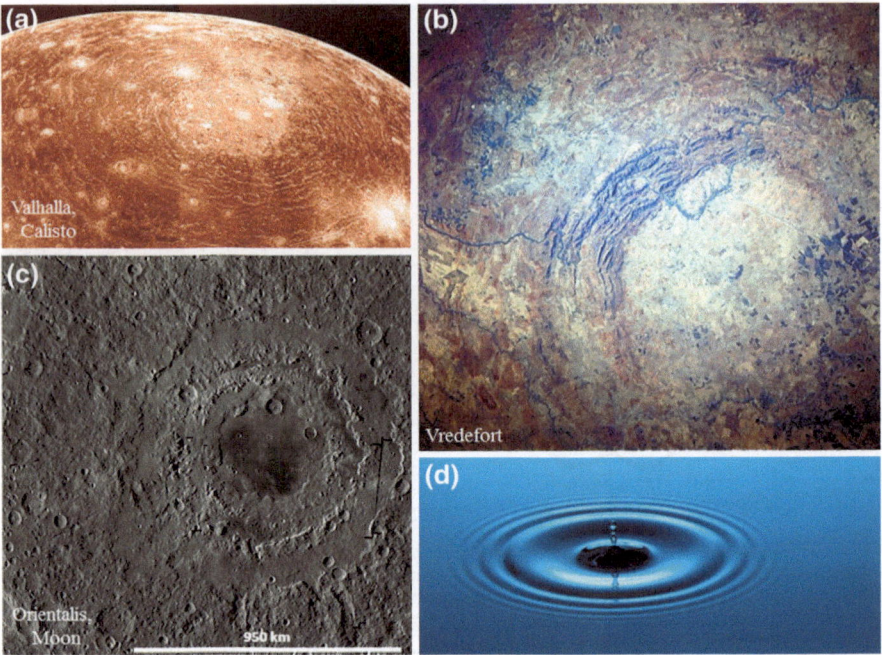

Fig. 4.3 Multi-ring impact structures. **a** Valhalla, Calisto—inner diameter of central region—360 km, outer ring <3,800 km. (NASA). **b** Orientalis, Moon—outer ring 950 km (NASA). **c** Vredefort, Orange Free State, South Africa—outer ring 298 km (EID, NASA). **d** A water drop ejected by elastic rebound—an analogy with an impact event (courtesy Mustafa Mincel)

a dome, such as for example Gosses Bluff (central Australia) (Figs. 1.1, 4.1) and Upheaval Dome (Utah) (Fig. 4.2). The relations between the diameter of the original transient crater and the final structure are estimated as 0.5–0.7 (Therriault et al. 1997). The formation of multiple rings in very large impact structures (>300 km-diameter) is an extension of the structural uplift process which extends to peak ring patterns and multi-ring structures (Grieve et al. 1981; Melosh 1989), such as are displayed by the ~900 km-diameter Orientale on the Moon (Fig. 4.3).

References

Dence MR (1968) Shock zoning at Canadian craters: Petrography and structural implications. In: French BM, Short NM (eds) Shock metamorphism of natural materials. Mono Book Corp Baltimore, pp 169–184

Dence MR, Grieve RAF, Robertson PB (1977) Terrestrial impact structures: principal characteristics and energy considerations. In: Roddy DJ, Pepin RO, Merrill RB (eds) Impact and explosion cratering: planetary and terrestrial implications. Pergamon New York, pp 247–275

French BM (1998) Traces of catastrophe—a handbook of shock metamorphic effects in terrestrial meteorite impact structures. Lunar Planet Sci Instit Contrib 954, p 120

Gault DE, Quaide WL, Oberbeck VR (1968) Impact cratering mechanics and structures. In: French BM, Short NM (eds) Shock metamorphism of natural materials. Mono Book Corp Baltimore, pp 87–99

Grieve RAF, Dence MR, Robertson PB (1977) Cratering process: as interpreted from the occurrence of impact melts. In: Roddy DJ, Pepin RO, Merrill RB (eds) Impact and Explosion Cratering: Planetary and Terrestrial Implications. Pergamon Press, New York, NY, pp 791–814

Grieve RAF, Cintala MJ (1992) An analysis of differential impact melt-crater scaling and implications for the terrestrial meteorite record. Meteoritics 27:526–538

Grieve RAF, Pilkington M (1996) The signature of terrestrial impacts. Aust Geol Surv J Aust Geol Geophys 16:399–420

Grieve RAF, Robertson PB, Dence MR (1981) Constraints on the formation of ring impact structures based on terrestrial data. In: Multi-ring Basins: formation and evolution. Proc Lunar Planet Sci 12A, Pergamon New York, pp 37–57

Holsapple KA, Schmidt RM (1982) On the scaling of crater dimensions 2: Impact processes. J Geophys Res 87:1849–1870

Holsapple KA, Schmidt RM (1987) Point source solutions and coupling parameters in cratering mechanics. J Geophys Res 92:6350–6376

Kieffer SW, Simonds CH (1980) The role of volatiles and lithology in the impact cratering process. Rev Geophys Space Phys 18:143–181

Melosh HJ (1989) Impact cratering: a Geological process. Oxford University Press, New York 245 p

Morrison D (2006) Asteroid and comet impacts: the ultimate environmental catastrophe. Phil Trans Roy Soc London 364:2041–2054. http://rsta.royalsocietypublishing.org/content/364/1845/2041/F1.expansion.html

O'Keefe JD, Ahrens TJ (1977) Impact-induced energy partitioning melting and vaporization on terrestrial planets. In: Proceedings of Lunar Science 8th Conference, pp 3357–3374

O'Keefe JD, Ahrens TJ (1993) Planetary cratering mechanics. J Geophys Res 98:17011–17028

Shoemaker EM (1963) Impact mechanics at Meteor Crater Arizona. In: Middlehurst BM, Kuiper GP (eds) The Moon Meteorites and Comets. University of Chicago, pp 301–336

Stöffler D, Bischoff L, Oskierski W, Wiest B (1988) Structural deformation breccia formation and shock metamorphism in the basement of complex terrestrial impact craters: implications for the cratering process. In: Bodén A, Eriksson KG (eds) Deep drilling in crystalline bedrock, vol 1. The deep gas drilling in the Siljan impact structure Sweden and astroblemes. Springer, New York, pp 277–297

Therriault AM, Grieve RAF, Reimold WU (1997) Original size of the Vredefort Structure: implications for thegeological evolution of the Witwatersrand Basin. Met Planet Sci 32:71–77

Chapter 5
Identification of Impact Structures

Abstract Confident criteria have been defined which allow the identification of high-velocity extraterrestrial impacts based structural morphometric parameters by surface mapping and geophysical exploration, microstructural shock metamorphic features corresponding to shock pressures >10 GPa, solid-state melting of target materials, mineralogical, geochemical and isotopic criteria.

5.1 Buried Impact Structures

The discovery by Robert Dietz (1914–1995) of large asteroid impact structures, referred to as 'astroblemes' (star scar), including Vredefort (298 km; 2,023 ± 4 Ma) (Dietz 1961), Sudbury (250 km; 1,850 ± 3 Ma) (Dietz 1964) and Gosses Bluff (24 km; 142.5 ± 0.8) (Dietz 1967; Milton et al. 1996) heralded a new era in the study of the asteroid impact history of Earth. Since then the advent of geophysical exploration and drilling uncovered a number of large buried impact structures identified by large circular gravity and magnetic anomalies and confirmed by diagnostic hallmarks of shock metamorphism. These discoveries included Chesapeake Bay (85 km; 35.5 ± 0.3 Ma) (Poag 1996), Chicxulub (170 km; 64.98 ± 0.05 Ma) (EID 2011) and Manson (35 km; 73.8 ± 0.3 Ma) (Koeberl and Anderson 1996). In Australia Woodleigh (Carnarvon Basin, Western Australia— 120 km to ~359 Ma); Glikson et al. 2005a, b), Tookoonooka (Eromanga Basin, Queensland—55–65 km; ~125 Ma); Gostin and Therriault 1997) and Talundilly (Eromanga Basin, Queensland ~84 km, 125 Ma; Gorter and Glikson 2012) impact structures were proven. In addition a number of probable and possible impact structures, including Gnargoo (north Carnarvon Basin, Western Australia; Iasky et al. 2001; Iasky and Glikson 2005), Mount Ashmore (Timor Sea; Glikson et al. 2010), and Warburton (northeast South Australia; Glikson and Uysal 2010, 2011; Glikson et al. 2013) display distinct structural and microstructural features diagnostic of impact and remain to be confirmed.

The discrimination between structural and microstructural features diagnostic of shock metamorphism and those of terrestrial origin has been the subject of extensive debate (Carter 1965, 1968; Carter and Friedman 1965; Carter et al. 1986; Alexopoulos et al. 1988; Lyons et al. 1993; Grieve et al. 1996, Vernooij and

A Y. Glikson, *The Asteroid Impact Connection of Planetary Evolution*,
SpringerBriefs in Earth Sciences, DOI: 10.1007/978-94-007-6328-9_5,
© The Author(s) 2013

Langenhorst 2005; Spray and Trepmann 2006; Ferrière et al. 2009; French and
Koeberl 2010; Hamers and Drury 2011). Central to the recognition of diagnostic
features of buried impacts are their unique geophysical characteristics, including
presence of a central uplift and a rim syncline, distinct circular and radial fault
patterns, microstructural shock metamorphic effects and extraterrestrial geochemical and isotopic signatures, distinct from structural features associated with
tectonic domes, salt domes or volcanic uplifts. Diagnostic criteria for definition
of extraterrestrial impacts are discussed below, with particular reference to confirmed, probable and possible buried Australian impact structures, using the wealth
of geophysical data from Australian sedimentary basins. Clarification of these criteria should help further investigation of seismic, magnetic and gravity data from
sedimentary terrains, with the aim of extending the impact data base, with implications for the effects of extraterrestrial impacts on crustal evolution.

5.2 Geophysical Criteria

Features suggestive of an impact origin of circular structures include intersections of older pre-impact structural elements by their external rings. These
are commonly but not invariably of near-complete 360°, as displayed by the
Chicxulub impact structure, Woodleigh impact structure and Gnargoo probable
impact structure (Fig. 5.1). Whereas structurally discordant intersections are also

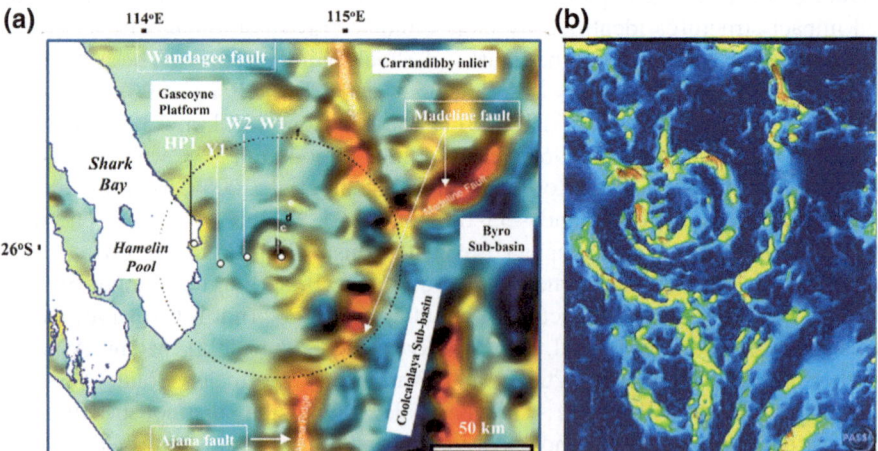

Fig. 5.1 Comparison between Bouguer anomaly signatures of Woodleigh and Chicxulub impact
structures. **a** First vertical derivative Bouguer anomaly image of the Woodleigh impact structure,
Gascoyne platform, southern Carnarvon basin, Western Australia. *W1*—Woodleigh 1A; *W2*—
Woodleigh 2A; *Y*—Yaringa 1 (Glikson et al. 2005a, b. AJES, Geological Society of Australia,
by permission). **b** Bouguer anomaly map of the Chicxulub impact structure, northern Yucatan
Peninsula (courtesy A. Hildebrandt and Earth Impact Database)

displayed by volcanic diatremes and by salt domes, the combination of features, including multi-ring patterns, high degree of circularity, presence of a central uplift core or central dome, and a ring syncline, is consistent with diagnostic features militating in favor of an impact origin.

The central uplifts of impact structures are best defined by seismic reflection sections where the basement uplift is commonly associated with thrust faults (Fig. 5.2). Where the core of the structure consists of sedimentary strata, a dome is outlined which may contain chaotically disrupted core zones displaying a loss of seismic markers due to mega-brecciation, as in the Mount Ashmore probable impact structure (Fig. 5.3). Whereas the central core zone is defined by thrust faults, the ring synclines and the outer rims of impact structures feature inward-dipping normal faults (Fig. 5.2). These structural patterns represent centripetal to upward block movements which involve compression around the uplifted core, extension within the ring syncline and inward collapse of the crater rim, as

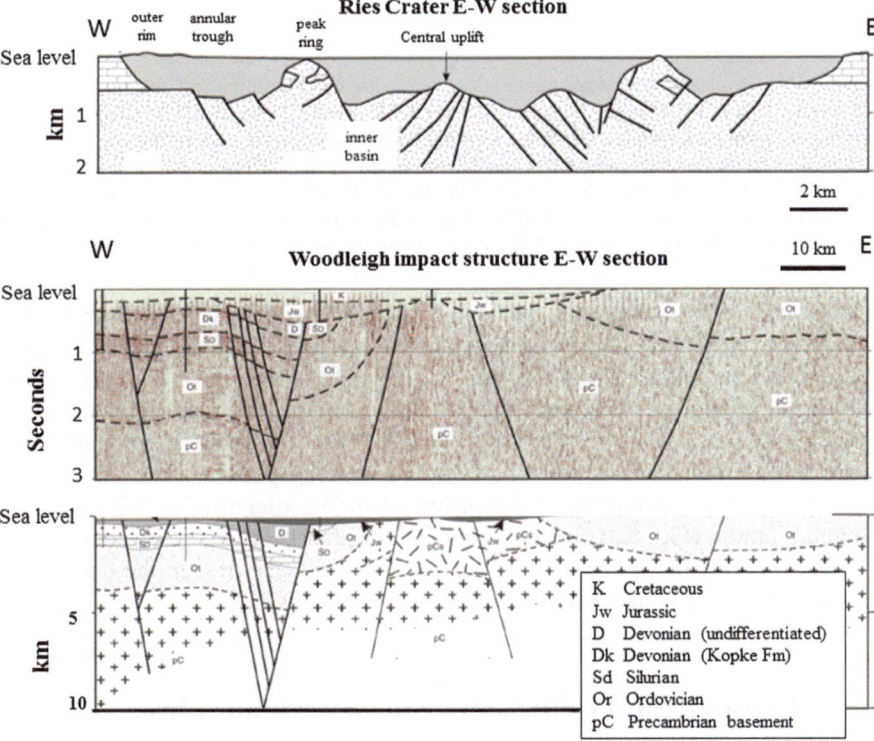

Fig. 5.2 Cross sections through the Ries impact structure, southern Germany (after Poag et al. 2004) and seismic-reflection profile and geological interpretation through the Woodleigh impact structure, Carnarvon Basin, Western Australia (Glikson et al. 2005a, b. AJES, Geological Society of Australia, by permission). Note the prevalence of thrust faults around the central uplifts, and of normal faults in the peripheries of both structures, representing centripetal and upward dislocation of blocks toward the centre

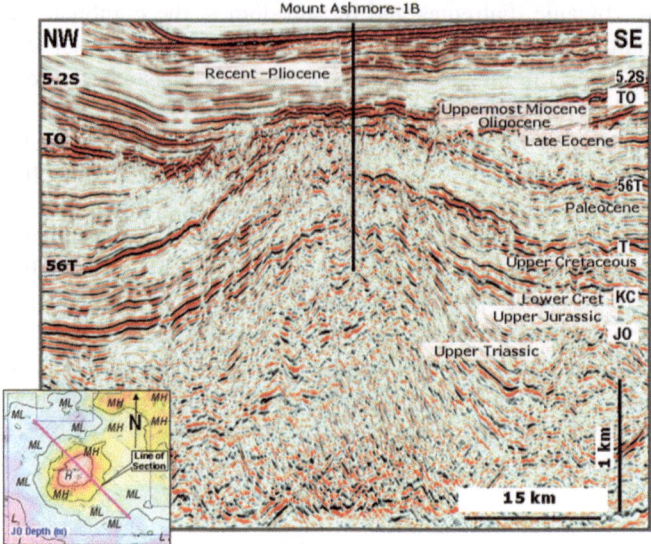

Fig. 5.3 Northwest–Southeast seismic section displaying block faulting and chaotic deformation in the inner core of the mount Ashmore structural dome, piercing of Cretaceous strata by upward movement of Jurassic and Triassic material, a marked base Oligocene unconformity and protrusion of the top of the dome above the pre-Eocene unconformity. The isopach map indicates the depth to JO (Oxfordian) (symbols in inset: *H*, high; *MH*, mid-high; *M*, mid; *ML*, mid-low; *L*, low. The location of the section is indicated by a line on the isopach map. Note vertical exaggeration. Glikson et al. 2010. AJES, Geological Society of Australia, by permission)

evident in the Woodleigh, Gnargoo and Talundilly structures. Some impact structures feature uplift of crystalline basement below the impacted sediments, as in the Woodleigh impact structure (Fig. 5.2). In other structures seismic reflection suggests an underlying basement plug, as in the Mount Ashmore probable impact structure (Glikson et al. 2010). Intersections of the top of uplifted sedimentary dome or basement plug by unconformably overlying post-impact sediments are typical of impact structures, as demonstrated in Woodleigh (Fig. 5.2), Gnargoo, Mount Ashmore (Fig. 5.3) and Tookoonooka. Further post impact isostatic vertical movements are indicated where the central uplift pierces through the unconformity, as for example in the Mount Ashmore structure (Fig. 5.3).

5.3 Microstructural Criteria for Shock Metamorphism

The distinction between Planar Deformation Features in quartz (Qz/PDF) indicative of shock metamorphism in established impact structures, on the one hand, and proposed Metamorphic Deformation Lamellae (Qz/MDL) of supposed purely endogenic origin, on the other hand, is extensively discussed in the literature (Carter 1965, 1968; Carter and Friedman 1965; Carter et al. 1986; Alexopoulos et al. 1988; Lyons et al.

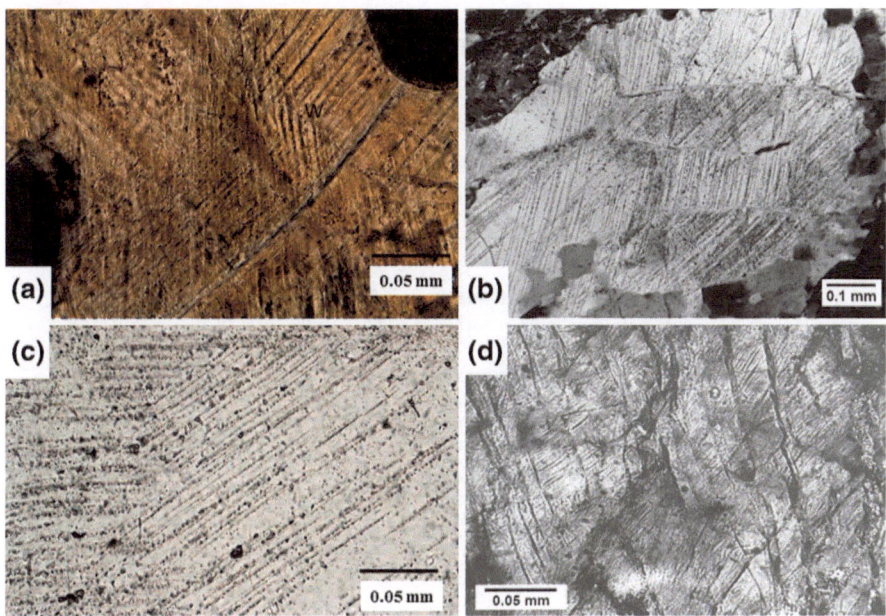

Fig. 5.4 Planar deformation features (PDF) diagnostic of shock metamorphism associated with high velocity impacts. **a** Woodleigh impact structure, Carnarvon Basin, Western Australia. **b** Multiple decorated (fluid inclusion-bearing) PDFs from an inclusion in suevite breccia from Rochechouart impact structure, France, showing two prominent sets of PDFs. Original partly continuous PDF traces are still recognizable (French 1998, by permission). **c** PDF from the Yarrabubba impact structure, Murchison Goldfields, Western Australia. **d** PDF in clastic quartz grains in Coconino Sandstone, Barringer Meteor Crater, Arizona (French 1998, by permission)

1993; Grieve et al. 1996; Vernooij and Langenhorst 2005; Spray and Trepmann 2006; Ferrière et al. 2009; French and Koeberl 2010; Hamers and Drury 2011). An origin by shock metamorphism is based on comparisons between experimentally deformed single quartz crystals and Qz/PDF from shocked rocks, for example from the Ries crater, which underwent little post-impact deformation (Engelhardt and Bertsch 1969), as in the examples shown in Fig. 5.4. Qz/PDF diagnostic of shock metamorphism display Crystallographic Miller Indices consistent with shock effects commonly in excess of 10 GPa (French, 1998) (Fig. 5.5). However, the presence of deformed quartz micro-structures within confirmed impact structures, such as Vredefort, Sudbury, Charlevoix and Manicouagan, where Qz/PDF orientations retain relics of shock-indicative Miller indices (Fig. 5.6), suggests the latter are derived by deformation of original Qz/PDF microstructures, as illustrated by the following examples:

1. In the Vredefort impact structure (Table 1.2; Sect. 9.2): Qz/PDF may be bent, clouded or decorated, and are dominated by basal {0001} sets (Grieve et al. 1990, Fig. 2) and to a lesser extent multiple sets (Fricke et al. 1990, Fig. 7), including quartz grains in clasts forming enclaves in granophyre veins (Buchanan and Reimold 2002, Fig. 5). Grieve et al. (1990, Fig. 2) compare the Vredefort planar

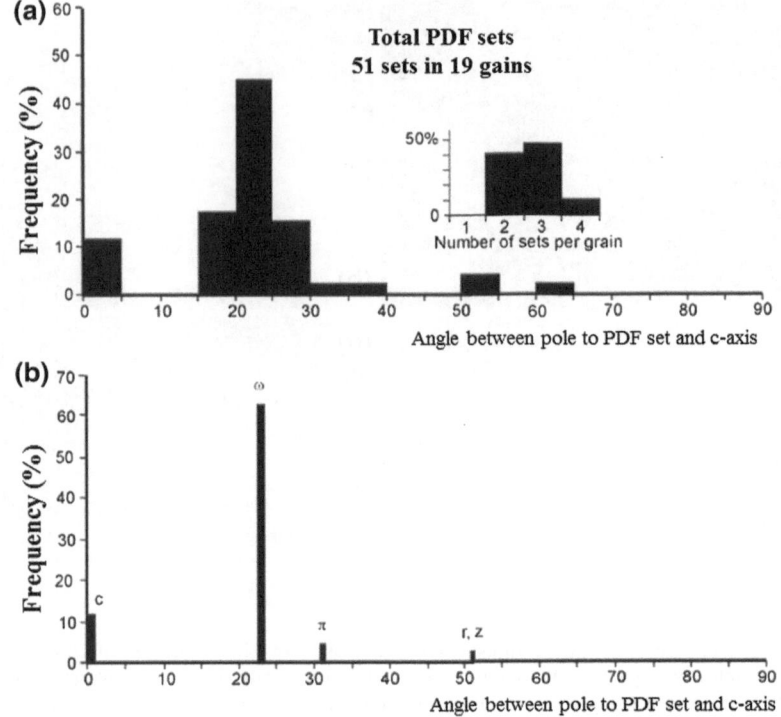

Fig. 5.5 PDF orientation in quartz in sample from the Woodleigh impact structure. **a** Frequency histogram of all PDF measurements plotted in 58 bins of angle between the pole to the set and the c-axis. Inset: frequency histogram of number of measured sets per grain. **b** Frequency histogram of indexed PDFs. Analyses by P. Haines (in Glikson et al. 2005a, b; AJES, Geological Society of Australia, by permission)

features with slightly undulating planar features from Mistatsin and Charlevoix impact structures. These authors conclude the planar features are "anomalous but still suggestive of an impact origin";

2. Charlevoix impact structure (Table 1.2), containing bent {0001} lamellae in quartz (Trepmann and Spray 2004, Fig. 5.6);
3. Manicouagan impact structure (Table 1.2) (Robertson 1975, Fig. 3a; Dressler 1990, Fig. 7);
4. Impact breccia of the Onaping Formation, Sudbury impact structure (Grieve et al. 2010, Figs. 4 and 6) (Sect. 9.3).

Criteria for distinction of planar deformation features include:

1. Qz/PDF lamellae are defined by diagnostic Miller indices correlated with specific shock levels, in particular the {10–11}, {10–12} and {10–13} planes (French 1998). By contrast planar features referred to as Qz/MDL show a wide scatter, including low $C_{OAQz}{}^\wedge P_{PE}$ angles of ~20–30° on frequency distribution plots (Lyons et al. 1993; French 1998, Fig. 4.25).

Fig. 5.6 Secondarily deformed planar deformation features (PDF) around established impact structures: **a** Charlevoix impact structure, Quebec (Trepmann and Spray 2004, by permission). **b** Yarrabubba impact structure, Murchison Goldfields, Western Australia

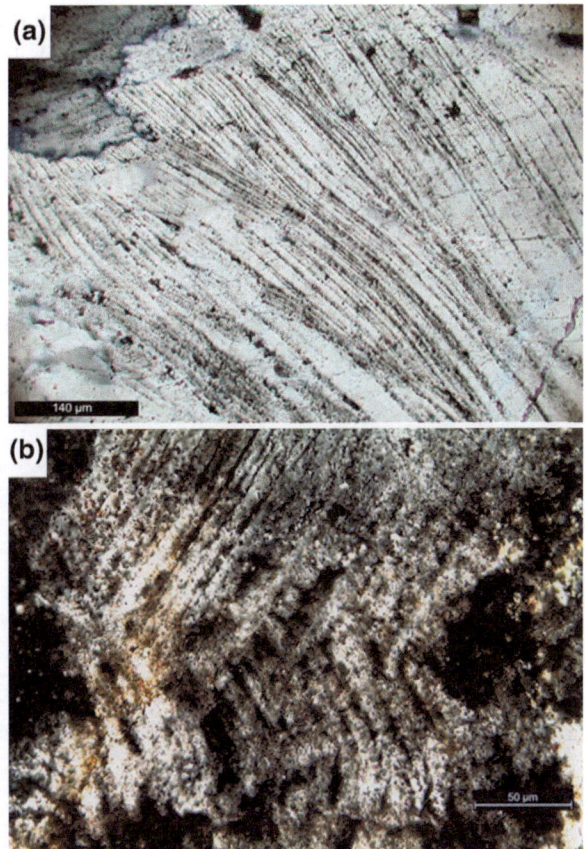

2. Qz/PDF form multiple planar sets in shock metamorphosed rocks (Robertson et al. 1968; Stoffler and Langenhorst 1994; Grieve et al. 1996). Planar features referred to as Qz/MDL mostly consist of only one set of lamellae within any one quartz grain (French and Koeberl 2010), commonly form basal {0001} planes which represent ~8–10 GPa shock pressures (French 1998, Table 4.2), evidenced in explosive volcanic units such as at Toba ignimbrites (Carter et al. 1986).
3. Qz/PDF lamellae can be <1–2 μm-thick whereas planar features referred to as Qz/MDL consist of segments usually 2–4 < μm thick.
4. Qz/PDFs are originally perfectly planar, but can be deformed, whereas planar features referred to as Qz/MDL commonly display undulation, bending and wavy patterns.
5. In transmission electron microscopy (TEM) little-deformed segments of Qz/PDFs display optical continuity between bordering crystal segments, namely they have no subgrain boundaries. By contrast planar features referred to as Qz/MDL display optical discontinuities between separate subgrains where optic orientations depart by ≤5° from each other and from the orientation of the host quartz (Glikson et al. 2013).

6. TEM studies indicate Qz/PDFs are either amorphous or composed of quartz with low dislocation densities. By contrast planar features described as Qz/MDL are more commonly altered and display high dislocation densities (Goltrant et al. 1991).
7. Qz/PDF in SEM–Cathode Luminescence form straight, narrow, well-defined features, whereas tectonic deformation lamellae are thicker and slightly curved (Hamers and Drury 2011).

Based on the above, criteria for distinguishing between Qz/PDFs associated with established impact structures and overprinted by ductile deformation, and criteria for planar features referred to as Qz/MDL of assumed endogenic tectonic-metamorphic origin, are yet to be agreed. The latter include Boehm lamellae {0001} documented from explosive volcanics, for example ignimbrites at Toba,

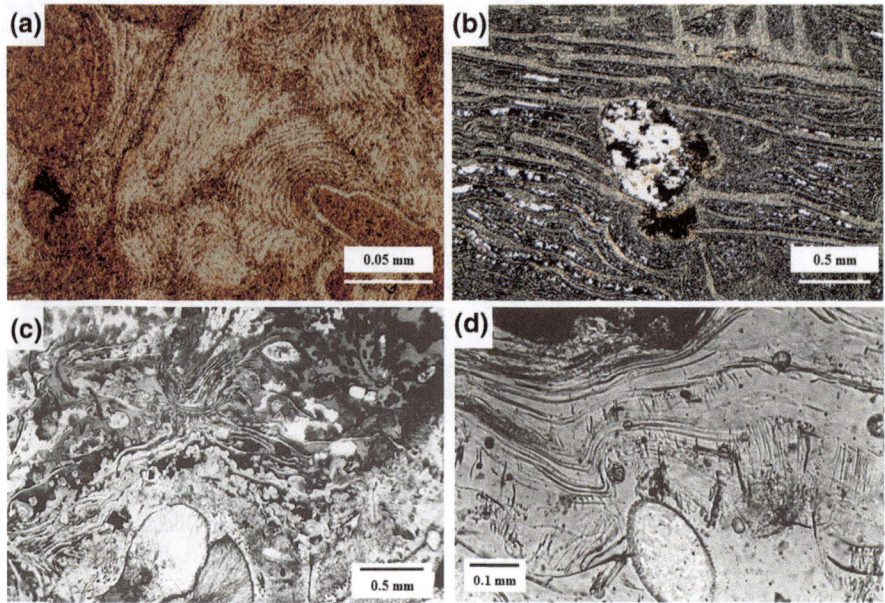

Fig. 5.7 Impact melts: **a** Flow-banded devitrified glass surrounding xenolith fragments of recrystallized granitoid, Yarrabubba impact structure, Western Australia. Plane polarized light. **b** Flow-banded devitrified glass surrounding xenolith fragments of recrystallized granitoid, Yarrabubba impact structure, Western Australia, in cross-nicols. **c** highly shocked, melted and recrystallized rock inclusion in metamorphosed suevite breccia. Post-shock temperatures apparently exceeded the melting points of all component minerals, converting the originally crystalline rock into an initially heterogeneous glass that developed limited flow textures before it was quenched. Onaping breccia, Sudbury impact structure. Plane polarized light (French 1998, by permission). **d** flow-banded glass from fragment in suevite breccia. Locally laminar flow-banding contains streaks and bands of quartz glass (Lechatelierite) (clear) formed by shock melting of original quartz grains at temperatures above 1,700 °C. Part of an included quartz grain (*dark*) appears at *top*, with flow-banding distorted around it. A filled vesicle (*white*) appears at bottom. West Clearwater Lake (Canada). Plane polarized light (French 1998, by permission)

Sumatra (Carter et al. 1986). Doubts remain regarding the origin of planar elements formed by processes other than impact-induced shock such as suggested by Lyons et al. (1993). Examples presented by these authors include: (1) Sandstones and quartzite from Finland and California showing either single or multiple crossing planar elements and (2) the Tapeat Sandstone which contains grains with multiple fine scale parallel lamella of ~4–5 μm width, with a scatter of 10°–55° on (C_{OAQZ}^P_{PDF}) frequency plots. Whether these quartz grains contain endogenic tectonic features or, alternatively, are related to or derived from an impacted source, remains unclear.

Vernooij and Langenhorst (2005) proposed criteria for discrimination between quartz lamella produced by shock metamorphism (Qz/PDF) and by tectonic deformation (Qz/MDL) in the following terms:

1. Qz/PDFs are perfectly planar instead of slightly curved (Fig. 14.1);
2. Qz/PDFs form strictly parallel, well defined, mostly rhombohedral crystallographic planes and Qz/PDF spacing is usually <1 μm instead of 2–4 μm;
3. Qz/PDFs do not induce a misorientation within the crystal and do not feature subgrain walls or dislocation bands, and
4. The density of free dislocations in Qz/PDF is low compared to Qz/MDL.

Fig. 5.8 The Popigai impact structure: **a** Photo showing megabreccia overlain by the crater melt-breccia sheet (Tagamite) (courtesy of Victor Masaitis). **b** A space view of Popigai (NASA). **c** Tagamite, Lapilli crystal-vitroclastic suevite. Angular clasts are mainly composed of chilled impact glass; matrix consists of partly altered small particles of the same glass and crystal fragments (Masaitis 2005, by permission)

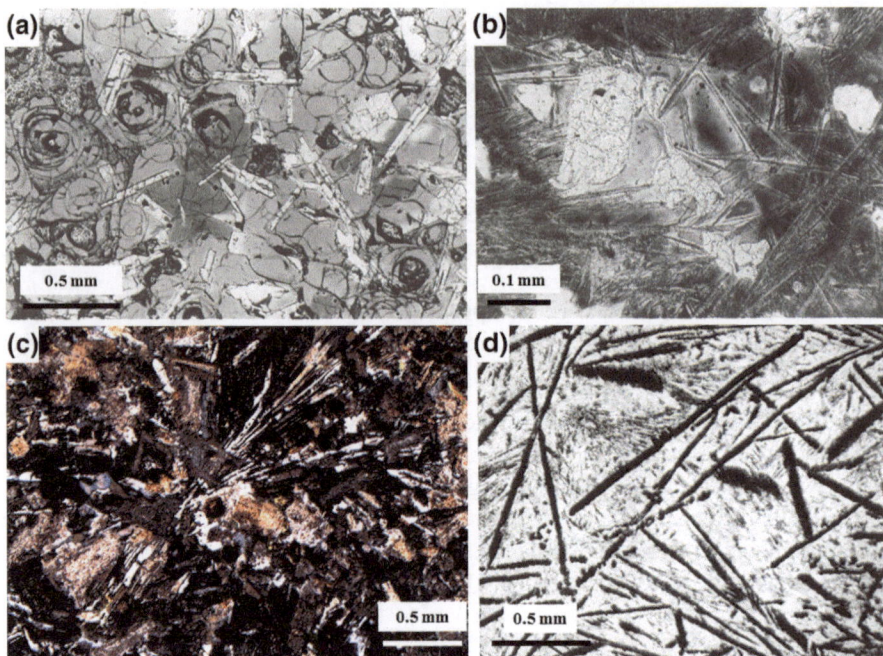

Fig. 5.9 Impact melts: **a** Glassy impact melt rock with euhedral quench feldspar needles in partly devitrified glassy matrix showing devitrification textures including circular cracks and spherulitic crystals. Identification as an impact melt was based on association with shock-metamorphosed rocks and the presence of anomalously high iridium in the melt rock. Lake Dellen, Sweden. Plane-polarized light (From French 1998, by permission). **b** Finely crystalline impact melt containing inclusions of quartz glass (Lechatelierite) in a partly crystalline matrix containing elongate quench pyroxene and clear interstitial brown glass. Lechatelierite displays a typical devitrification features. Tenoumer Crater (Mauritania). Plane-polarized light (from French 1998). **c** Radiating granophyre from the Yarrabubba impact structure, Murchison Goldfields, Western Australia, displaying radiating albite interspersed with devitrified brown glass and commonly surrounding cores of small xenoliths of remnant granite or feldspar crystals. **d** Quench crystallized orthopyroxene needles in spherulitic groundmass of intergrown feldspar and quartz in impact melt, forming a "Bronzite Granophyre" dike cutting pre-impact basement granites. Farm Koppieskraal, Vredefort structure, South Africa. Plane polarized light (From French 1998, by permission)

Rarely have multiple planar sets in quartz been documented from terrains not known to be associated with impact structures. French (1990), investigating impact suggestions for the layered Bushveld Complex, South Africa (Hamilton 1970; Elston and Twist 1986), analyzed planar microstructures in quartz from quartzite inclusions in the Rooiberg felsite, which display multiple lamellae. Lamella orientations display little correspondence to peak orientations corresponding to shock metamorphism-related Miller indices. The Rooiberg Felsite has been subject to little or no post-magmatic tectonic deformation. At the present state of knowledge an impact origin cannot be ruled out for the Bushveld Complex, rendering the quartz microstructures hardly a type locality for Qz/MDL.

At higher degrees of shock the micron-scale Qz/PDF lamella evolve into dia-plectic glass and melt fractions, which may contain relic shocked rock and mineral fragments containing Qz/PDF and feldspar/PDF lamella (Fig. 5.7). Melt fractions of crater breccia occur in numerous structures, including Brent Crater (Canada) and Tenoumer Crater (Mauritania). Quench crystallization around these fragments may result in fan-shaped crystallites of feldspar, quartz and chlorite, evolving into igne-ous–like textures and granophyre such as in 'Bronzite granophyre ' dykes of the Vredefort impact structure, Mistasin Lake impact structure and Yarrabubba impact structure (Fig. 5.7). In large impact structures segregation leads to formation of large impact melt sheets, examples being the Sudbury Igneous Complex (Therriault et al. 2002) (Sect. 9.3), Popigai impact melt (Tagamite) (Masaitis 1998) (Fig. 5.8), Mistastin Lake impact (French 1998), West Clearwater (Canada), Morokweng melt sheet (Hart et al. 2002) and the Yarrabubba granophyre plug (Macdonald et al. 2003) (Fig. 5.9). The Sudbury and Vredefort impact structures include offset dikes injected into the sub-crater basement (Wood and Spray 1998).

References

Alexopoulos JS, Grieve RAF, Robertson PB (1988) Microscopic lamellar deformation features in quartz: discriminative characteristics of shock-generated varieties. Geology 16:796–799

Buchanan PC, Reimold WU (2002) Planar deformation features and impact glass in inclusions from the Vredefort granophyre South Africa. Meteor Planet Sci 37:807–822

Carter NJ (1965) Basal quartz deformation lamellae—a criterion for recognition of impactites. Am J Sci 263:786–806

Carter NJ (1968) Meteoritic impact and deformation of quartz. Science 160:526–528

Carter NJ, Friedman M (1965) Dynamic analysis of deformed quartz and calcite from the Dry Creek Ridge Anticline Montana. Am J Sci 263:747–785

Carter NL, Officer CB, Chesnerc A, Rose WI (1986) Dynamic deformation of volcanic ejecta from the Toba caldera: possible relevance to Cretaceous/Tertiary boundary phenomena. Geology 14:380–383

Dietz RS (1961) Vredefort ring structure: meteorite impact scar? J Geol 69:496–505

Dietz RS (1964) Sudbury structure as an astroblemes. J Geol 72:412–434

Dietz RS (1967) Shatter cone orientation at Gosses Bluff astrobleme. Nature 216:1082–1084

Dressler BO (1990) Geochemistry of the impact-generated melt sheet at Manicouagan: evidence for fractional crystallization. J Geophys Res 116:22

EID—Earth Impact Database (2011) http://wwwpasscnet/EarthImpactDatabase/indexhtml

Elston WE, Twist D (1986) Bushveld complex South Africa: is Rooiberg felsite an impact melt? Lunar Planet Sci 17:204–205

Engelhardt WB, Bertsch I (1969) Shock induced planar deformation structures in quartz from the Ries Crater, Germany. Contr Mineral Petrol 20:203–234

Ferriere L, Morrow JR, Amgaa T, Koeberl C (2009) Systematic study of universal-stage measure-ments of planar deformation features in shocked quartz: implications for statistical signifi-cance and representation of results. Meteor Planet Sci 44:925–940

French BM (1990) Absence of shock metamorphic effects in the bushveld complex South Africa: results of an intensive search. Tectonophysics 171:287–301

French BM (1998) Traces of catastrophe—a handbook of shock metamorphic effects in terres-trial meteorite impact structures. Lunar Planet Sci Inst Contrib 954:120

French BM, Koeberl C (2010) The convincing identification of terrestrial meteorite impact struc-tures: what works what doesn't and why. Earth Sci Rev 98:123–170

Fricke A, Frick M, Medenbach O, Schreyer W (1990) Fluid inclusions planar elements and pseu-
dotachylites in the basement rocks of the Vredefort structure South Africa. Tectonophysics
171:169–183

Glikson AY, Uysal IT (2010) Evidence of impact shock metamorphism in basement granitoids,
Cooper Basin. In: Australian geothermal conference. Adelaide

Glikson AY, Uysal IT (2011) Geophysical anomalies and quartz microstructures of the East
Warburton Basin under the Cooper Basin South Australia: tectonic or asteroid impact origin?
http://wwwuqeduau/geothermal/growing-evidence-for-a-large-asteroid-hitting-cooper-basin-
300-million-years-ago-110557

Glikson AY, Eggins S, Golding S, Haines P, Iasky RP, Mernagh TP, Mory AJ, Pirajno F, Uysal
IT (2005a) Microchemistry and microstructures of hydrothermally altered shock-metamor-
phosed basement gneiss, Woodleigh impact structure, Southern Carnarvon Basin, Western
Australia. Aust J Earth Sci 52:555–573

Glikson A, Mory AJ, Iasky R, Pirajno F, Golding S, Uysal IT (2005b) Woodleigh, Southern
Carnarvon Basin, Western Australia: history of discovery late Devonian age and geophysical and morphometric evidence for a 120 km-diameter impact structure. Aust J Earth Sci
52:545–553

Glikson AY, Jablonski D, Westlake S (2010) Origin of the mount Ashmore structural dome West
Bonaparte Basin Timor Sea. Aust J Earth Sci 57:411–430

Glikson AY, Uysal IT, Fitz Gerald JD, Saygin E (2013) Geophysical anomalies and quartz microstructures, Eastern Warburton Basin, North-east South Australia: Tectonic or impact shock
metamorphic origin? Tectonophysics 589:57–76

Goltrant O, Cordier P, Doukhan JC (1991) Planar deformation features in shocked quartz: a
transmission electron microscopy investigation. Earth Planet Sci Lett 106:103–115

Gorter JD, Glikson AY (2012) Talundilly Western Queensland Australia: geophysical and petrological evidence for an 84 km-large impact structure and an early Cretaceous impact cluster.
Aust J Earth Sci 59:51–73

Gostin VA, Therriault AM (1997) Tookoonooka: a large buried early Cretaceous impact structure
in the Eromanga Basin of Southwestern Queensland Australia. Meteor Planet Sci 32:593–599

Grieve RAF, Corderre JM, Robertson PB, Alexopuolos J (1990) Microscopic planar deformation
features in quartz of the Vredefort structure: anomalous but still suggestive of an impact origin. Tectonophysics 171:185–200

Grieve RAF, Langenhorst F, Stoffler D (1996) Shock metamorphism of quartz in nature and
experiment: II significance in geoscience. Metor Planet Sci 31:6–35

Grieve RAF, Ames DE, Morgan JV, Artmieva N (2010) The evolution of the Onaping formation
at the Sudbury impact structure. Meteor Planet Sci 45:159–782

Hamers MF, Drury MR (2011) Scanning electron microscope cathod-luminescence (SEM-CL)
imaging of planar deformation features and tectonic deformation lamellae in quartz. Meteor
Planet Sci 46:1814

Hamilton WB (1970) Bushveld complex-product of impacts? Geol Soc S Afr Sp Publ 1:367–379

Hart RJ, Cloete M, McDonald I, Carlson RW, Andreoli MAG (2002) Siderophile-rich inclusions
from the Morokweng impact melt sheet, South Africa: possible fragments of a chondritic
meteorite. Earth Planet Sci Lett 198:49–62

Iasky RP, Glikson AY (2005) Gnargoo: a possible 75 km-diameter post-early Permian—pre-
Cretaceous buried impact structure Carnarvon Basin Western Australia. Aust J Earth Sci
52:577–586

Iasky RP, Mory AJ, Blundell KA (2001) The geophysical interpretation of the Woodleigh impact
structure Southern Carnarvon Basin. Western Australia, Geol Surv of West Aust Rep 79

Koeberl C, Anderson RR (1996) The manson impact structure Iowa: anatomy of an impact crater.
Geol Soc of Am Pap 302:468

Lyons JB, Officer CB, Borella PE, Lahodynsky R (1993) Planar lamellar substructures in quartz.
Earth Planet Sci Lett 119:431–440

Macdonald FA, Bunting JA, Cina SE (2003) Yarrabubba—a large deeply eroded impact structure
in the Yilgarn Craton Western Australia. Earth Planet Sci Lett 213:235–247

Masaitis VL (1998) Popigai crater: origin and distribution of diamond-bearing impactites. Meteor Planet Sci 33:349–359

Masaitis VL (2005) Morphological, structural and lithological records of terrestrial impacts: an overview. Aust J Earth Sci 52:509–528

Milton DJ, Barlow BC, Brown AR, Moss FJ, Manwaring EA, Sedmik ECE, Young GA, Van Son J (1996) Gosses Bluff—a latest Jurassic impact structure central Australia: part 2- seismic magmatic and gravity studies. Aust Geol Surv Org J Aust Geol Geophys 16:487–527

Poag CW (1996) Structural outer rim of Chesapeake Bay impact crater: seismic and bore hole evidence. Meteor Planet Sci 31:218–226

Poag CW, Koeberl C, Reimold WU (2004) The Chesapeake Bay crater: geology and geophysics of a late Eocene submarine impact structure. Springer, Berlin, p 522

Robertson PB (1975) Zones of shock metamorphism at the Charlevoix impact structure, Quebec. Bull Geol Soc Am 86:1630–1638

Robertson PB, Dence MR, Vos MA (1968) Deformation in rock-forming minerals from Canadian craters. In: French BM, Short NM (eds) Shock metamorphism of natural materials. Mono Book Corp Baltimore MD433–452

Spray JG, Trepmann CA (2006) Shock-induced crystal-plastic deformation and post-shock annealing of quartz. European J Mineral 18:161–173

Stoffler D, Langenhorst F (1994) Shock metamorphism of quartz in nature and experiment: I basic observation and theory. Meteoritics 29:155–181

Therriault AM, Anthony D, Flower R, Grieve RAF (2002) The Sudbury igneous complex: a differentiated impact melt sheet. Econ Geol 97:1521–1540

Trepmann C, Spray JG (2004) Post-shock crystal plastic processes in quartz from crystalline target rocks of the Charlevoix impact structure. Lunar Planet Sci 35:1370

Vernooij MJC, Langenhorst F (2005) Experimental reproduction of tectonic deformation lamella in quartz and comparison to shock-induced planar deformation features. Meteor Planet Sci 40:1353–1361

Wood CR, Spray JG (1998) Origin and emplacement of offset dykes in the Sudbury impact structure: constraints from Hess. Meteor Planet Sci 33:337–347

Chapter 6
Impact Ejecta and Fallout Units

Abstract The discovery of altered glass spherules (microkrystites) formed by atmospheric condensation of impact-released vapor at the Cretaceous-Tertiary (K-T) impact boundary by Alvarez et al. (Science 208:1095–1108, 1980) opened the way to the identification of impact ejecta units in Archaean and Proterozoic terrains and thereby investigation of the impact history of the early Earth.

Large extraterrestrial impacts ensue in a wide range of atmospheric, hydrospheric, seismic, and tectonic consequences, including tsunami waves, turbidity currents, sea-floor turbulence, slumping, faulting, detachment and transport of blocks derived from fault scarps, the classic case being tsunami deposits associated with the Cretaceous-Tertiary (KT) boundary Chicxulub crater (Alvarez et al. 1980, 1982; Smit and Klaver 1981; Hildebrand and Boynton 1990; Hildebrand et al. 1991; Smit et al. 1992, 1994a, b; Kelley and Gurov 2002; Scheffers and Kelletat 2003). A late Precambrian example is the Acraman impact structure, South Australia, and the adjoining ejecta and fallout and tsunami units in the Flinders Ranges and Officer Basin (Gostin et al. 1989, 2010; Wallace et al. 1990, 1996).

The identification of impact ejecta and fallout units depends critically on the presence of microtektites and microkrystites. Microkrystite spherules are produced by condensation of impact-ejected vapor released when large bolides collide with Earth. On impact, target materials are shattered and brecciated, and the core surrounding the exploding bolide is evaporated. The crust and mantle underlying exploding projectile rebound elastically to form a dome (Grieve and Dence 1979; Grieve and Shoemaker 1994; Grieve and Pesonen 1996; Grieve and Pilkington 1996; Shoemaker and Shoemaker 1996; French 1998). The impact vapor is dispersed in the atmosphere, transported by winds, cools and condenses as a myriad melt droplets solidified as tiny glass spheres preserved in submarine sediments (Glass and Burns 1988; Melosh and Vickery 1991) (Figs. 1.4, 6.1). Microkrystite spherules are distinguished from volcanic melt droplets formed by volcanic fountaining by virtue of their inward-radiating quench crystallites and centrally offset vesicles, evidence of aerodynamic imbalance. By contrast volcanic fountain-produced glass spheres display outward-radiating quench crystallites.

Microkrystites may contain nickel-rich spinel, rare micrometer-scale metallic nuggets of nickel and platinum group elements, and geochemical and isotopic

A. Y. Glikson, *The Asteroid Impact Connection of Planetary Evolution*,
SpringerBriefs in Earth Sciences, DOI: 10.1007/978-94-007-6328-9_6,
© The Author(s) 2013

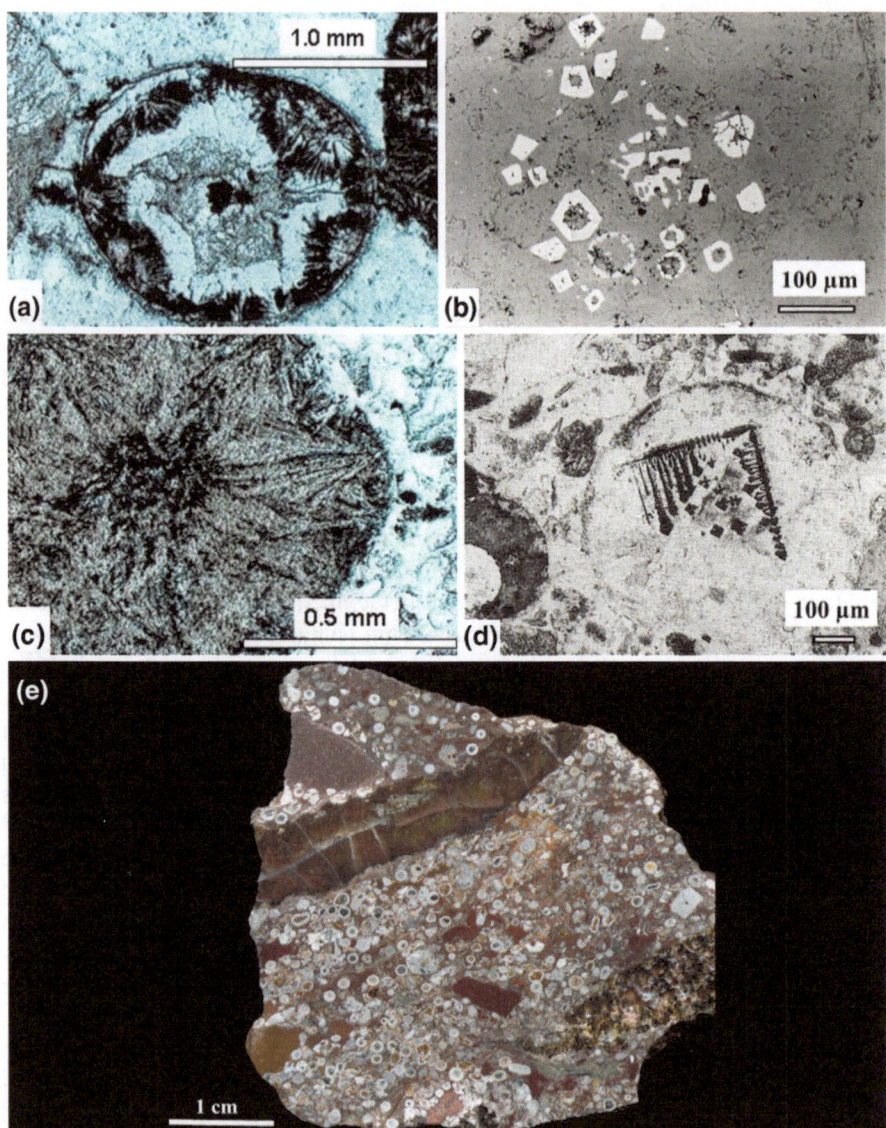

Fig. 6.1 S3 impact layer spherules, Barberton greenstone belt (BGB), Kaapvaal craton, South Africa. **a** Microkrystite spherule showing inward-radiating chlorite crystallites, likely pseudomorphs after quench pyroxene. **b** Detail of part of an S3 microkrystite spherule, showing inward-radiating chlorite fans and a central cavity. **c** Cluster of Ni-chromites crystals located within an S3 microkrystite spherule showing partial alteration by chlorite of the cores of spinels and along crystallographic planes. Plane polarized light (from Glikson 2005a, b). **d** Microkrystite spherule containing quench skeletal Ni-chromite. Spherule on lower left shows an off-center vesicle typical of microkrystite spherules (plane-polarized light) (courtesy of G.R. Byerly). **e** Sample of the Jeerinah impact layer, showing microkrystites with white K-feldspar shells and commonly chlorite-rich cores, agglomerated microkrystite spherules, possible microtektites, fragments of rip-up siltstone and a detached layer of chert (courtesy Bruce Simonson) (**c** and **d** by permission of AJES, Geological Society of Australia)

signatures including platinum group element anomalies with high iridium levels, high Ir/Pd ratios, and $^{53}Cr/^{52}Cr$ isotope anomalies (Byerly and Lowe 1994;
Shukolyukov et al. 2000; Simonson and Glass 2004; Glikson 2005a, b). Iridium,
whose chondritic abundance is between 2 and 3 orders of magnitude higher than
in the Earth's mantle (Fig. 7.1), provides the a key parameter, used by Alvarez
et al. (1982) to establish the impact connection of the K–T boundary at 65 Ma.

Impact ejecta units are closely associated with tsunami deposits, which range
from cross-beds to fragmented material and large erratic boulders and debris
flow to in situ disrupted submarine beds. In the case of the ~2.63 Ga Carawine
Dolomite, where a spherule-bearing mega-breccia 10–30 meter-thick extends over
a distance of more than 100 km (Simonson and Hassler 1997) (Fig. 8.6), the tsunami wave affected pelagic (below wave-base) carbonates, testifying to possibly
over 200 m-deep wave amplitudes affecting the sea bed (Glikson 2004).

The identification of impact ejecta and fallout units in the field is complicated
by the millimeter-scale of microkrystites (Glass and Burns 1988), microtektites,
and meta-glass shards (Figs. 1.5, 6.1). As distinct from Phanerozoic sequences,
where mass extinction boundaries may yield clues for abrupt events, the scarcity of
fossils in the Precambrian record hardly allows such approach. On the other hand,
the lack of bioturbation in Precambrian sediments results in superior preservation
of the fine-scale textures of impact fallout deposits (Simonson and Harnik 2000).

The distinction between impact-triggered tsunami deposits and tsunami deposits unrelated to impacts hinges on the presence of diagnostic extraterrestrial components, including microkrystites and microtektites (Smit and Klaver 1981; Glass
and Burns 1988; Izett et al. 1990; Claeys et al. 1992; Smit et al. 1992; Wang 1992;
Glass and Wu 1993) (Fig. 1.5), Ni-rich chromite spinels (Kyte and Smit 1986;
Robin et al. 1991; Byerly and Lowe 1994) (Fig. 6.1), shocked quartz (Figs. 5.4,
5.6) (Bohor et al. 1984), platinum group element (PGE) enrichment (Alvarez et
al. 1980, 1982) (Fig. 7.1) PGE ratios (Kyte et al. 1980; Glikson and Allen 2004)
(Fig. 7.1), chondritic Cr isotopic ratios (Shukolyukov et al. 2000; Kyte et al. 2003)
(Fig. 7.3), Osmium isotopic ratios (Luck and Turekian 1983), Rhodium (Bekov
et al. 1988), nanophase Fe-rich material (Wdowiak et al. 2001), fallout stishovite
(Bohor et al. 1984; McHone et al. 1989), and fallout shocked zircon and fallout
diamonds (Carlisle and Braman 1991). Other suggested, but to date not confirmed,
criteria for extraterrestrial contributions include Helium anomalies (Farley et al.
1998) (Fig. 7.4) and ^3He in fullerenes (Heymann et al. 1994; Becker et al. 2000)
and amino acids showing extraterrestrial symmetry (Zhao and Bada 1989).

Impact fallout units typically contain rip-up clasts at their base, signifying
impact-triggered seismic disturbance of the sea floor prior to spherule settling and
the arrival of a tsunami (Fig. 1.5a). Impact-triggered tsunami diamictite may preserve microkrystites and microtektites, otherwise destroyed by winnowing in shallow water and reef environments. Impact fallout spherules may in turn overlain by
diamictite and erratic boulders signifying deep amplitude storm or tsunami effects
(Fig. 8.5c, d). Because of the abrupt high-energy effects of a tsunami, below-normal
wave base effects and subsequent burial, little reworking by currents may follow.

Depending on current intensities, seismic and/or impact-generated seismic
effects are represented by the occurrence of high-energy sedimentary units within

pelagic below-wave base sequences, including (1) turbidity current deposition
of immature arenites which display cyclic graded cycles (Simonson 1992) (Fig.
8.8c); (2) presence of tsunami-transported exotic fragments and boulders within
pelagic sequences (Fig. 8.10); (3) cross-bedded, climbing cross-bedded, and
turbulence-induced arenite eddies within siltstones; (4) sedimentary overfolds
and slumps, probably representing current switching and/or slumping of liquefied muds over steep slopes; (5) excavation and brecciation of solid substratum,
accompanied with injection of spherule-bearing liquefied muds induced by tsunami waves (Fig. 8.7); and (6) occurrence of pebble- to boulder-size fragments
within spherule layers attributable to fault scarp detachment and tsunami transport
(Fig. 8.10).

Tsunami deposits associated with impact ejecta and fallout microkrystite and
microtektite units in the Pilbara Craton, Western Australia, include, from older
to younger: (1) chert intraclast-bearing arenite and diamictite associated with
microkrystite spherules in the Antarctic Creek Chert Member (ACM), lower
Mount Ada Basalt (3.47 Ga), central Pilbara (Lowe and Byerly 1986; Byerly
et al. 2002) (Fig. 8.2); (2) boulder mass flow conglomerate overlying the pre-
2629 ± 5 Ma Jeerinah Impact Layer (JIL) (Simonson et al. 2000a, b, c, 2001)
(Fig. 8.5); (3) Spherule Marker Bed (SMB)-1 and SMB-2 of the 2561 ± 8 Ma
(Trendall et al. 1998) Wittenoom Formation; (4) excavated mega-breccia of
the Carawine Dolomite, possibly contemporaneous with the SMB (Simonson
1992); and (5) Erratic boulders of chert, BIF and carbonate in the 2481 ± 4 Ma
Dales Gorge Shale-4 Macroband (DGS4) impact fallout unit (Hassler and
Simonson 2001) (Fig. 8.9). These studies document an association between
impact fallout units and current-disrupted and debris flow deposits located in
below-wave base, located within sequences of pelagic sediments dominated
by siltstone, carbonate, chert, and banded iron formation. Current-perturbed
deposits typically comprise arenite, turbidite, rip-up clast-bearing basal spherule layers, spherule-rich intraclast conglomerate, and erratic boulders, marked
by climbing ripples, turbidite eddies, and slump structures. The sharp contrast
between the low-energy host deep-water sediments and the high-energy deposits
is interpreted in terms of deep-amplitude tsunami waves (Hassler et al. 2000;
Glikson 2004).

References

Alvarez L, Alvarez W, Asaro F, Michel HV (1980) Extraterrestrial cause for the Cretaceous-
    Tertiary extinction. Science 208:1095–1108
Alvarez L, Alvarez W, Asaro F, Michel HV (1982) Iridium anomaly approximately synchronous
    with terminal Eocene extinctions. Science 216:886–888
Becker L, Poreda R, Bunch T (2000) Fullerenes: an extraterrestrial carbon carrier phase for noble
    gases. Proc Natl Acad Sci 97:2979–2983
Bekov GI, Letokhov VS, Radaev VN, Badyukov DD, Nazarov MA (1988) Rhodium distribution
    at the Cretaceous/Tertiary boundary analyzed by ultrasensitive laser photoionization. Nature
    332:146–148

Bohor BF, Foord EE, Modreski PJ, Triplehorn DM (1984) Mineralogic evidence for an impact event at the Cretaceous-Tertiary boundary. Science 224:867–869

Byerly GR, Lowe DR (1994) Spinels from Archaean impact spherules. Geochim et Cosmochim Acta 58:3469–3486

Byerly GR, Lowe DR, Wooden JL, Xie X (2002) A meteorite impact layer 3470 Ma from the Pilbara and Kaapvaal cratons. Science 297:1325–1327

Carlisle DB, Braman DR (1991) Nanometre-size diamonds in the Cretaceous-Tertiary boundary clay of Alberta. Nature 352:708–709

Claeys P, Casier JG, Margolis SV (1992) Microtektites and mass extinctions evidence for a late Devonian asteroid impact. Science 257:1102–1104

Farley KA, Montanari A, Hoemaker EM, Shoemaker C (1998) Geochemical evidence for a comet shower in the Late Eocene. Science 280:1250–1253

French BM (1998) Traces of catastrophe—a handbook of shock metamorphic effects in terrestrial meteorite impact structures. Lunar Planet Sci Inst Contrib 954:120

Glass BP, Burns CA (1988) Microkrystites: a new term for impact-produced glassy spherules containing primary crystallites. In: Proceedings of lunar planet science conference, vol 18. pp 455–458

Glass BP, Wu J (1993) Coesite and shocked quartz discovered in the Australasian and North American microtektite layers. Geology 21:435–438

Glikson AY (2004) Bedout: a possible end-permian impact crater offshore of northwestern Australia. Science 306:613

Glikson AY (2005a) Geochemical and isotopic signatures of Archaean to early Proterozoic extraterrestrial impact ejecta/fallout units. Aust J Earth Sci 52:785–799

Glikson AY (2005b) Geochemical signatures of Archaean to early Proterozoic Mare-scale oceanic impact basins. Geology 133:125–128

Glikson AY, Allen C (2004) Iridium anomalies and fractionated siderophile element patterns in impact ejecta, Brockman Iron Formation, Hamersley Basin, Western Australia: evidence for a major asteroid impact in *simatic* crustal regions of the early Proterozoic earth. Earth Planet Sci Lett 20:247–264

Gostin VA, Keays RR, Wallace MW (1989) Iridium anomaly from the Acraman ejecta horizon: impacts can produce sedimentary iridium peaks. Nature 340:542–544

Gostin VA, McKirdy DM, Webster LJ, Williams GE (2010) Ediacaran ice-rafting and coeval asteroid impact, South Australia: insights into the terminal Proterozoic environment. Aust J Earth Sci 57(7):859–869

Grieve RAF, Dence MR (1979) The terrestrial cratering record: II the crater production rate. Icarus 38:230–242

Grieve RAF, Pesonen LJ (1996) Terrestrial impact craters: their spatial and temporal distribution and impacting bodies. Earth Moon Planets 72:357–376

Grieve RAF, Pilkington M (1996) The signature of terrestrial impacts. Aust Geol Surv J Aust Geol Geophys 16:399–420

Grieve RAF, Shoemaker EM (1994) The record of past impacts on Earth. University of Arizona Press, Tucson, pp 417–462

Hassler SW, Robey HF, Simonson BM (2000) Bedforms produced by impact-generated tsunami, ~2.6 Ga Hamersley Basin, Western Australia. Geology 135:283–294

Heymann D, Chibante LPF, Brooks RR, Wolbach WS, Smalley RE (1994) Fullerenes in the Cretaceous-Tertiary boundary layer. Science 265:645–647

Hildebrand A, Boynton WV (1990) Proximal Cretaceous-Tertiary boundary impact deposits in the Caribbean. Science 248:843–847

Hildebrand AR, Penfield GT, Kring DA, Pilkington M, Camargo ZA, Jacobsen SB, Boynton WV (1991) A possible Cretaceous-Tertiary boundary impact crater on the Yucatan Peninsula, Mexico. Geology 19:867–871

Izett GA, Maurrasse FJ-MR, Lichte FE, Meeker GP, Bates R (1990) Tektites in Cretaceous-Tertiary boundary rocks on Haiti. US Geological Survey Open-File Report 90–635

Kelley SP, Gurov E (2002) The Boltysh another end-Cretaceous impact. Meteor Planet Sci 37:1031–1044

Kyte FT, Smit J (1986) Regional variations in spinel compositions: an important key to the Cretaceous/Tertiary event. Geology 14:485–487

Kyte FT, Zhou Z, Wasson JT (1980) Siderophile enriched sediments from the Cretaceous-Tertiary boundary. Nature 288:651–656

Kyte FT, Shukolyukov A, Lugmair GW, Lowe DR, Byerly GR (2003) Early Archaean spherule beds: chromium isotopes confirm origin through multiple impacts of projectiles of carbonaceous chondrite type. Geology 31:283–286

Lowe DR, Byerly GR (1986) Early Archean silicate spherules of probable impact origin South Africa and Western Australia. Geology 14:83–86

Luck JM, Turekian KK (1983) 187-osmium-186/osmium in manganese nodules and the Cretaceous-Tertiary boundary. Science 222:613–615

McHone JF, Nieman RA, Lewis CF, Yates AM (1989) Stishovite at the Cretaceous-Tertiary boundary, Raton, New Mexico. Science 243:1182–1184

Melosh HJ, Vickery AM (1991) Melt droplet formation in energetic impact events. Nature 350:494–497

Robin E, Boclet D, Bonte P, Froget L, Jehanno C, Rocchia R (1991) The stratigraphic distribution of Ni-rich spinels in Cretaceous-Tertiary boundary rocks at El-Kef (Tunisia) Caravaca, Spain, and Hole-761C (Leg-122). Earth Planet Sci Lett 107:715–721

Scheffers A, Kelleat D (2002) Sedimentologic and geomorphologic tsunami imprints worldwide—a review. Earth Sci Rev 63:83–92

Shoemaker EM, Shoemaker CS (1996) The Proterozoic impact record of Australia. Aust Geol Surv Org J Aust Geol Geophys 16:379–398

Shukolyukov A, Kyte FT, Lugmair GW, Lowe DR, Byerly GR (2000) The oldest impact deposits on Earth. In: Koeberl C, Gilmour I (eds) Lecture notes in Earth science 92: impacts and the early Earth. Springer, Berlin, pp 99–116

Simonson BM (1992) Geological evidence for an early Precambrian microtektite strewn field in the Hamersley Basin of Western Australia. Geol Soc Am Bull 104:829–839

Simonson BM, Glass BP (2004) Spherule layers—records of ancient impacts. Ann Rev Earth Planet Sci 32:329–361

Simonson BM, Harnik P (2000) Have distal impact ejecta changed through geologic time? Geology 28:975–978

Simonson BM, Hassler SW (1997) Revised correlations in the early Precambrian Hamersley Basin based on a horizon of resedimented impact spherules. Aust J Earth Sci 44:37–48

Simonson BM, Davies D, Hassler SW (2000a) Discovery of a layer of probable impact melt spherules in the late Archaean Jeerinah Formation, Fortescue Group, Western Australia. Aust J Earth Sci 47:315–325

Simonson BM, Hornstein M, Hassler SW (2000b) Particles in late Archean Carawine Dolomite, Western Australia, resemble Muong Nong-type tektites. In: Gilmour I, Koeberl C (eds) Impacts and the early earth. Springer, Berlin, pp 181–214

Simonson BM, Koeberl C, McDonald I, Reimold WU (2000c) Geochemical evidence for an impact origin for a late Archean spherule layer Transvaal supergroup South Africa. Geology 28:1103–1106

Simonson BM, Cardiff M, Schubel KA (2001) New evidence that a spherule layer in the late Archaean Jeerinah Formation of Western Australia was produced by a major impact. In: 32nd lunar planetary science conference abstracts, lunar and planetary institute contribution 1080, Houston

Smit J, Klaver G (1981) Sanidine spherules at the Cretaceous-Tertiary boundary indicate a large impact event. Nature 292:47–49

Smit J, Montanari A, Swinburne NHM, Alvarez W, Hildebrand AR, Margolis SV, Claeys P, Lowrie W, Asaro F (1992) Tektite-bearing deep-water clastic unit at the Cretaceous-Tertiary boundary in northeastern Mexico. Geology 20:99–103

Smit J, Roep TB, Alvarez W, Claeys P, Montanari A (1994a) Deposition of channel deposits near the Cretaceous-Tertiary boundary in northeastern Mexico catastrophic or normal sedimentary deposits and is there evidence for Cretaceous-Tertiary boundary-age deep-water deposits in the Caribbean and Gulf of Mexico?: comment. Geology 22:953–954

Smit J, Roep TB, Alvarez W, Claeys P, Montanari S, Grajales M (1994b) Impact tsunami-generated clastic beds at the KT boundary of the Gulf coastal plain A synthesis of old and new outcrops. In: New developments regarding the KT event and other catastrophes in Earth history. Lunar Planet Inst Houston Contrib 825:117–119

Trendall AF, Nelson DR, deLaeter JR, Hassler SW (1998) Precise zircon U-Pb ages from the Marra Mamba Iron Formation and Wittenoom Formation, Hamersley Group, Western Australia. Aust J Earth Sci 45:137–142

Wallace MW, Gostin VA, Keays RR (1990) Spherules and shard-like clasts from the late Proterozoic Acraman impact ejecta horizon South Australia. Meteoritics 25:161–165

Wallace MW, Gostin VA, Keays RR (1996) Sedimentology of the Neoproterozoic Acraman impact-ejecta horizon South Australia. Aust Geol Surv Org J Aust Geol Geophys 16:443–451

Wang K (1992) Glassy microspherules (microtektites) from an upper Devonian limestone. Science 256:1547–1550

Wdowiak TJ, Armendarez LP, Agresti DG, Wade ML, Wdowiak SY, Claeys P, Izett G (2001) Presence of an iron-rich nanophase material in the upper layer of the Cretaceous-Tertiary boundary clay. Meteor Planet Sci 36:123–133

Zhao M, Bada JL (1989) Extraterrestrial amino acids in Cretaceous/Tertiary boundary sediments at Stevns Klint, Denmark. Nature 339:463–465

Chapter 7
Extraterrestrial Geochemical, Isotopic and Mineralogical Signatures

Abstract The unique geochemical and isotopic characteristics of asteroids, including Platinum Group Elements (PGE—Ir, Rh, Pt, Pd) abundances and ratios, $^{53}Cr/^{52}Cr$ isotopes, $^{182}W/^{183}W$ isotopes, $^{187}Os/^{188}Os$ isotopes and other parameters, as compared to terrestrial materials, and the presence of Ni and PGE nanonuggets and Ni-rich Cr spinels, allow confident identification of extraterrestrial contributions to melts, ejecta and sediments.

7.1 Geochemical Signatures

The geochemical, isotopic and mineralogical information provided by impact ejecta and fallout units allows insights into the composition of the impact target crust and the parent projectile, as well as a first approximation of the size of the projectiles and thereby of the ensuing craters. Impact ejecta and fallout units commonly display elevated abundances of Nickel, Cobalt and Chromium, reflecting an extraterrestrial component derived from the parental projectiles. As such enrichment in ferromagnesian elements can also result from involvement of terrestrial mafic and ultramafic igneous rocks and derived concentrates, such high abundances do not in themselves provide diagnostic criteria for an extraterrestrial contribution. However, as Ni/Cr ratios vary by a factor of about 4–5 between chondritic (Ni/Cr ~4) and mantle materials (Ni/Cr <1.0), high Ni/Cr ratios in sediments potentially provides initial clues for an extraterrestrial component (ETC). Some microkrystite-rich fallout units display Ni, Co and Cr abundances and ratios similar or, in some instances, higher than komatiites. For example chlorite-dominated microkrystites of the BGB-S3 fallout unit have Ni/Cr ratios similar to, or exceeding those of komatiites. More highly altered and carbonated impactites tend to have lower siderophile levels, with Ni–Cr–Co relations similar to those of sediments and the mean upper Archaean crustal value.

In view of the two to three orders of magnitude difference between the abundance of Iridium in chondrites (~450 ppb) and mantle pyrolite (~3.2 ppb) (MacDonough and Sun 1995) (Fig. 7.1), this element allows more confident approximations of the ETC in impact ejecta units. Absolute levels and ratios of the platinum group elements provide essential information on: (1) composition of the

A. Y. Glikson, *The Asteroid Impact Connection of Planetary Evolution*,
SpringerBriefs in Earth Sciences, DOI: 10.1007/978-94-007-6328-9_7,

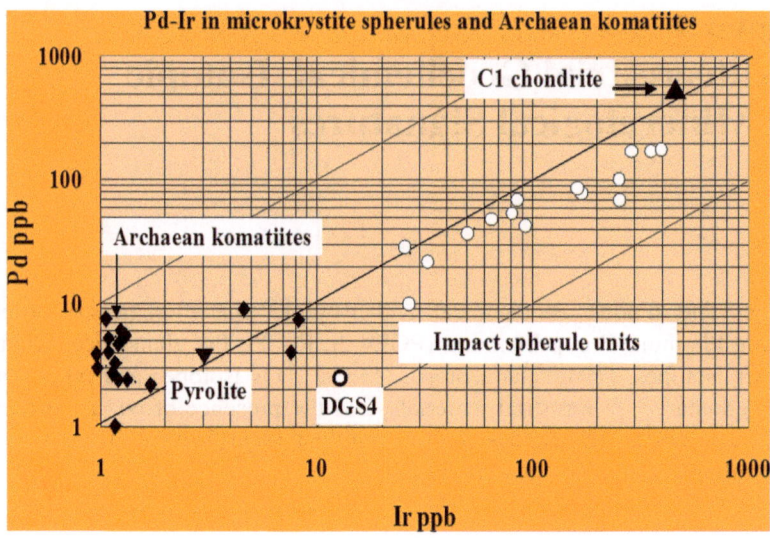

Fig. 7.1 Concentrations of Iridium and palladium in Archaean komatiite basalts, mantle pyrolite, microkrystite spherules of the Barberton Greenstone belt (BGB) and the DGS4 unit (Hamersley Basin) and C1-chondrites, displaying distinct fields and ratios where impact materials have Pd/Ir ratios of ≥1.0

projectile; (2) relative contribution of meteoritic and terrestrial components to the impact ejecta, and (3) first-order approximation of the size of the parental asteroid by mass-balance calculations (Byerly and Lowe 1994; Shukolyukov et al. 2000; Kyte et al. 2003; Glikson and Allen 2004) (Fig. 7.2).

PGE patterns of microkrystite spherule-bearing impactites in the Pilbara and Kaapvaal Cratons display consistent differences from those of volcanic and sedimentary rocks. The Iridium abundances of impactite samples are mostly several ppb to a few tens of ppb, whereas volcanics (komatiites—mostly <2 ppb) and sediments (mostly <0.1 ppb) are at least one order of magnitude less. Impactites are relatively enriched in refractory PGEs (Rh, Ru, Os, Ir) and depleted in relatively volatile PGEs (Au, Pd, Pt), showing low Pd/Ir ratios (Fig. 7.1). These relations are contrasted with those of volatile PGE-enriched volcanic and sedimentary rocks. Secondary alteration of impactites results in an increase in the volatile/refractory PGE ratios, as indicated by the following examples: (1) high Pd/Ir ratios in carbonated spherules of the ~2.56 Ga Spherule Marker Bed; (2) high Pd/Ir ratios in altered zones of the ~0.58 Ga Bunyeroo ejecta (Wallace et al. 1990) and (3) sharp increase in Au and in the Au/Ir and Pd/Ir ratios associated with sulfide and gold mineralization of BGB-S3 impactites of the Barberton Greenstone Belt as contrasted with non-mineralized impactites (Reimold et al. 2000).

Excepting impact fallout units, volatile PGE-depleted terrestrial rocks include refractory harzburgites (Tredoux et al. 1989). Given the significant differences in boiling points of the PGEs (Au—2850 °C; Pd—3000 °C; Pt—3720 °C; Rh—3970 °C Ir—4550 °C), the loss in volatile PGE species may be interpreted

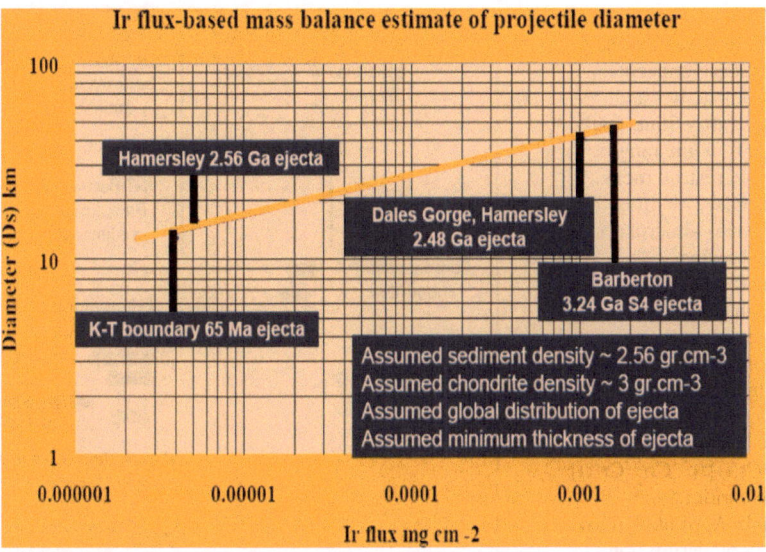

Fig. 7.2 Correlation between Ir flux (in units of 10^{-4} mg cm^{-2}) and diameter of chondritic projectile (Dp), based on mass-balance calculations assuming mean unit thickness, mean Ir concentration, and global distribution of ejecta. Unit symbols: Rp (projectile radius) = $\sqrt{[Vp/(4/3)\pi]}$; Vp = Mp/dp; Mp = FG/Cp; FG$_{Earth}$ = AE/dS × TS × S$_{Earth}$ (Earth's surface area), where AE is measured element (E) abundance in ejecta unit (in ppb); dS is mean density of ejected materials (mg/cm^3) (assumed as 2.55 g/cm^3); CE = mass of element E (in mg/cm3); (CE = AE × DS); TS = mean stratigraphic thickness of spherule unit (in mm); FE = local mean flux of element E (in mg per cm^2 surface; FE = CE × TS); FG = inferred global flux (in mg/cm^2) of element E; Cp = assumed concentration of element E in projectile (ppb) (for C1 chondrites with 450 ppb Ir); Mp = mass of projectile; dp = assumed density of projectile (C1 chondrites' density = 3.0 g/cm^3)

in terms of preferential condensation of the refractory PGE (Glikson and Allen 2004). Since Pd and Pt abundances are commonly high relative to Ir and therefore more readily analyzed by ICPMS, whereas Ir may occur near or below detection limits, Pd/Pt ratios allow practical geochemical tracer for discrimination of impact fallout materials. Byerly and Lowe (1994), on the basis of the bulk composition of the Barberton BGB-S3 and BGB-S4 impactite layers (Al_2O_3—12.5 %; TiO_2—0.75 %; Cr—1000 ppm; Zr—40 ppm; Ir—78 ppb), deduced a composition consisting of target components (basalt, komatiite, minor dacite), with an added chondrite component.

7.2   Isotopic Signatures

Shukolyukov et al. (2000) suggested that negative $^{53}Cr/^{52}Cr$ ($\varepsilon^{53}Cr$) values of the Barberton BGB-S4 microkrystite spherule-rich unit are consistent with derivation from carbonaceous chondrites, as contrasted with ordinary chondrites and planetary

Fig. 7.3 εCr values of several meteorite classes, K–T boundary impactites, BGB-S4 impactites and sediments (after Shukolyukov et al. 2000). *White* range *arrows*—terrestrial rocks and minerals; *grey*-filled range *arrows*—microkrystite spherule-bearing impactites; *black* range *arrows*—meteorites and impactites. Note the correspondence between impactite ε^{53}Cr range and carbonaceous chondrites. ε^{53}Cr = $10^4\{^{53}$Cr/^{52}C(m)—$(^{53}$Cr/^{52}Cr(t)$\}$/$\{^{53}$Cr/^{52}Cr(t)$\}$ (m—meteoritic; t—terrestrial). A. Shukolyukov and AJES by permission

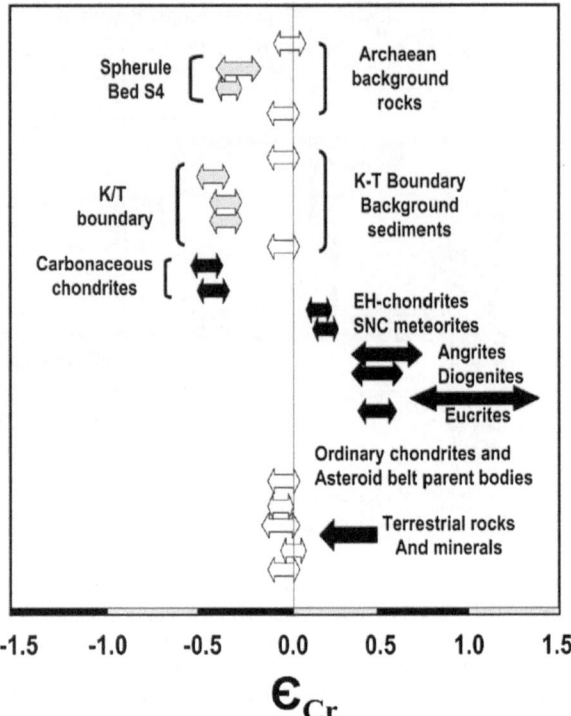

bodies with positive ε^{53}Cr values, consistent with suggestions by Kyte et al. (2003) (Fig. 7.3). The extraterrestrial component is estimated by these authors from mass balance of Ir and Cr and from the proportion of ε^{53}Cr. On this basis the spherule beds contain a high mean extraterrestrial component (BGB-S2 ~0.4–26 %; BGB-S3 ~46–100 %; BGB-S4 ~15–80 %: Kyte et al. 2003 Table 2). Ir-based mass-balance estimates for the BGB projectiles, about two orders of magnitude higher than those for the K–T boundary impact, yield estimated diameters of ~30 km in diameter (Sleep et al. 1989; Kyte et al. 1992; Byerly and Lowe 1994) (Fig. 7.2). Estimates based on spherule size distribution (Melosh and Vickery 1991) yield similar large diameters, although such estimates contain larger uncertainties as compared to mass-balance calculations. Kyte et al. (2003) estimated projectiles with diameters ~3–7 times the Chicxulub projectile (~10 km diameter). Similar results are obtained from mass-balance estimates based on the proportion of ε^{53}Cr in the BGB impactites, suggesting diameters of >20 km for parental projectiles of BGB-S4 and yet larger diameter for BGB-S2 and S3 (Shukolyukov et al. 2000). Mass-balance estimates of projectile diameters based on Ir and Pt abundances in Pilbara microkrystite-rich fallout units are given in Glikson and Allen (2004), including estimates for the JIL (~2.63 Ga), SMB-1 (2.56 Ga) and DGS4 (~2.48 Ga) ejecta units. These estimates suggest a ~2–3 % extraterrestrial component in DGS4. The inferred diameters of the JIL projectile (~20 km) and DGS4 projectile (~18 km) are on a similar scale as the

Barberton projectiles. These estimates depend critically on assumed mean global ejecta thicknesses and are therefore subject to large errors of no less than $\pm$ 50 %.

The largely ferromagnesian composition of microkrystite spherules, indicated by high Ni, Co and Cr abundances, is interpreted in terms of mafic crustal targets (Simonson et al. 1998; Glikson and Allen 2004). Information on crustal or mantle sources of host rocks of the microkrystites can be furnished by $^{147}Sm/^{144}Nd$, $^{176}Lu/^{177}Hf$ and $^{87}Rb/^{86}Sr$ isotopic systems. The distribution with time of initial $\varepsilon^{143}Nd$ (McCulloch and Bennett 1994), $\varepsilon^{176}Hf$ (Bizzarro et al. 2003) and initial $^{87}Sr/^{86}Sr$ values (Glikson 1983) suggest a dominance of mantle-derived crust during the Archaean, in agreement with the abundance of mafic and ultramafic relics in Early Archaean gneiss terrains.

The radiogenic decay of short-lived isotopes such as $^{53}Mn/^{53}Cr$ (half-life 3.7×10^6 years) and $^{182}Hf/^{182}W$ (half-life 9×10^6 years) in the early Solar System results in distinct isotopic anomalies between (1) meteoritic components which fractionated during the early solar period and remained isolated isotopic systems, and (2) mixed isotopic composition homogenized by mantle convection during accretion in the first $\sim 60 \cdot 10^6$ years of terrestrial evolution (Lugmair and Shukolyukov 1998; Schoenberg et al. 2002). High-precision mass spectrometry can resolve distinct Cr and W isotope ratios between different classes of meteorites that originate from fractionation of short-lived parent to daughter isotope. Systematic differences are observed between the $\varepsilon^{53}Cr$ values[1] of terrestrial sediments and extraterrestrial impact fallout units, including K–T boundary microkrystite-rich sediments (Shukolyukov and Lugmair 1998) and Archaean spherule-rich BGB-S4 and BGB-S3 spherule units (Shukolyukov et al. 2000; Kyte et al. 2003) (Fig. 6.1). The $\varepsilon^{53}Cr$ value of the analyzed BGB-S4 sample was determined as $0.32 + 0.06$, similar to $\varepsilon^{53}Cr$ values of K–T boundary spherule-bearing sediments and carbonaceous chondrites (Shukolyukov et al. 2000). Similar $\varepsilon^{53}Cr$ values were obtained for the BGB-S2 and BGB-S3 fallout units (Kyte et al. 2003). Analogous observations pertain to $\varepsilon^{182}W$ value[2] systematics where the chondritic and iron meteorite ratios are consistently depleted relative to the silicate Earth. The early mantle and crust display Hf enrichment relative to W likely due to early fractionation of tungsten into the core. A study of $\sim$ 3.8–3.7 Ga meta-sediments from southwest Greenland and Labrador by Schoenberg et al. (2002) indicates $\varepsilon^{182}W$ values as low as 1.37, consistent with the values of ordinary chondrites, indicating a meteoritic component.

The decay of ^{187}Re to ^{187}Os and marked increase in the Re/Os ratio upon magmatic fractionation renders the $\varepsilon^{187}Os$ value[3] diagnostic of Rhenium-depleted mantle products (abyssal peridotite $e^{187}Os = 0.125$) and meteoritic materials ($\varepsilon^{187}Os \sim 0.1286$) (Turekian 1982; Scheffers and Kelleat 2002). The Earth's mantle and chondritic meteorites had similar $\varepsilon^{187}Os$ values throughout geological history. By contrast, typical continental materials have higher $\varepsilon^{187}Os$ values (41.0), which

---

[1] $\varepsilon^{53}Cr = 10^4 \{^{53}Cr/^{52}Cr(m)—^{53}Cr/^{52}Cr(t)\}/\{^{53}Cr/^{52}Cr(t)\}$, m—meteoritic; t—terrestrial.

[2] $\varepsilon^{182}W = 10^4\{^{182}W/^{183}W(m)—^{182}W/^{183}W(t)\}/\{^{182}W/^{183}W(t)\}$, m—meteoritic; t—terrestrial.

[3] $\varepsilon Os = 10^4\{^{187}Os/^{188}Os(m)—^{187}O/^{188}Os(t)\}/\{^{187}Os/^{188}Os(t)\}$, m, meteoritic; t, terrestrial.

Fig. 7.4 Extraterrestrial ^{3}He flux (10^{12}cc STP cm^{-2} ka^{-1}) displayed by the Late Eocene section at Massignano, Macerata, Italy (from Farley et al. 1998, by permission and AJES by permission). (STP i.e. 1 atmosphere, 0 °C)

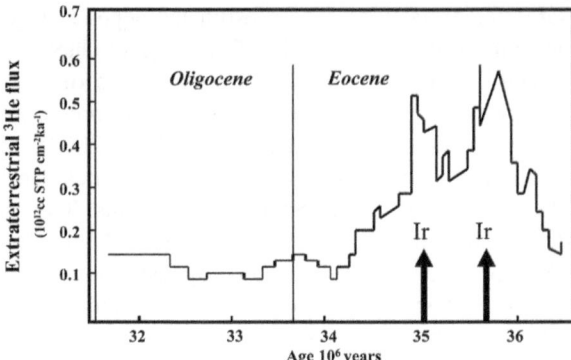

allows the identification of probable extraterrestrial components, as for example in the Vredefort granophyre dykes (e^{187}Os ~ 0.196–0.668) (Koeberl et al. 1996). However, the overlap of the chondritic and mantle fields renders Os isotopes somewhat less diagnostic than Cr and W isotopes. Meteoritic Pb isotopes, mostly below ^{206}Pb/^{204}Pb ~ 10 and ^{207}Pb/^{204}Pb ~ 10 in iron meteorites or troilites of low U/Pb ratios, may not be expected to be preserved in impact fallout units, due to the mobility of Pb and the extreme mobility of U. Oxygen isotopes, which display systematic differences between chondritic, planetary and terrestrial δ^{18}O (% rel. SMOW) to δ^{17}O (% rel. SMOW) ratios, are unlikely to be well preserved in the hydrous terrestrial environment of impact fallout units. However, despite alteration, remarkably meteoritic ^{3}He/^{4}He signatures (Farley et al. 1998) and racemic organic molecules (AIB—aminoisobutyric acid: Zhao and Bada 1989; Zahnle and Grinspoon 1990) have been recorded in K–T and E–O boundary (Fig. 7.4) impact fallout units. Such cometary 'seeding' are observed to both precede and post-date impact events, as recorded by Ir anomalies, a relationship yet to be tested in Precambrian sediments where higher degrees of alteration may obscure He and AIB signatures.

7.3 Mineralogical Signatures

Commonly well preserved inward-radiating fan-textured K-feldspar or chlorite crystallites, likely representing primary quench crystallization and/or devitrification features, dominate microkrystite spherules (Fig. 1.4) in ~2.63 Ga, ~2.56 Ga and ~2.48 Ga impactites in the Pilbara Craton. K-feldspar shells of stilpnomelane-dominated microkrystite spherules of the Dales Gorge DGS4 Macroband contains sub-micrometer-scale Ni particles, including Ni-metal, Ni-oxide, Ni–sulfide, Ni arsenide and Ni –Mg sulfide (Glikson and Allen 2004) (Fig. 8.9). The preservation of Ni nano-nuggets inclusions in quench K-feldspar shells suggests they represent relic original components of the microkrystites, that alteration was incomplete and the inward-radiating K-feldspar crystallites represent quench and/or devitrification features rather than diagenesis and hydrous burial metamorphism. Secondary iron oxides, sericite, chlorite, stilpnomelane and recrystallized K-feldspar may contain relic siderophile element patterns inherited from recrystallized glass. Relic textures are obliterated where microkrystites are replaced by carbonate and quartz. Octahedral and quench skeletal crystals of Ni-chromites occur in microkrystite spherules from the Cretaceous–Tertiary (K–T) boundary (Kyte and Smit 1986; Robin et al. 1991), the Eocene–Oligocene (E–O) boundary (Montanari et al. 1993), Pliocene Eltanin oceanic impact (Margolis et al. 1991) and the Barberton BGB-S3 impact fallout unit (Byerly and Lowe 1994). The spinels display unique chemistry (NiO <23 %; CoO <0.53 %; V2O3 <2.7 %; ZnO <3.25 %). Byerly and Lowe (1994) classified the Ni-chromites in terms of Al-bearing lithophile type and Al-poor chalcophile type, likely reflecting different mixtures between projectile and target materials. Low Fe^{+3}/Fe^{+2} ratios shown by the Barberton Greenstone Belt spherules, as compared to K–T spherules, have been interpreted in terms of low oxygen fugacities of the Archaean atmosphere (Byerly and Lowe 1994).

The meteoritic impact-related derivation of the spherules and their contained Ni-chromites was questioned by Koeberl et al. (2003) who pointed to (1) alteration of PGEs related to sulfide alteration and (2) possible derivation of the chromites by erosion of exposed nickel-rich intrusive ultramafic rocks of the Bon Accord type (Tredoux et al. 1989). However, the meteoritic impact connection of Ni-chromite in the spherules is confirmed by their distinct Ni values (<23 %) as compared to chromites of komatiites (NiO ~ 0.12 %) or chromites segregated by sulfide immiscibility (NiO ~ 0.26 %) (Czamanske et al. 1976). Bon Accord chromites do not occur as quench skeletal crystals within spherules, have lower Ni values and display enriched volatile PGE levels.

Kyte et al. (1992) reported an iridium nano-nugget within sulfides of the Barberton BGB-S4 spherule unit. laser-ICPMS analysis of Ni-chromites from BGB-S3 unit indicates PGE values below detection limit; however, the chromites contain PGE rich nano-nuggets (Glikson 2007). This suggests that in part the PGEs reside in Ni-chromites and may be preserved in sulfides that replace the chromite in mineralized sectors of the spherule units, such as are described in the Princeton and Sheba mines (Reimold et al. 2000). Microkrystite-bearing units commonly contain Ni-bearing magnetite grains and Nickel and Silica-bearing magnetite grains, analyzed in the JIL and DGS4 impact fallout units. Such mixed compositions have been reported in connection with meteoritic and possible meteoritic fallout units from the K–T and Permian—Triassic boundaries (Miura et al. 2001) and may result from metal–silicate reactions in the impact-ejected vapor cloud or/and from recrystallization of iron oxide and silica within Nickel-rich sedimentary matrix of the impactite.

References

Bizzaro M, Baker JA, Haack H, Ulfbeck D, Minik R (2003) Early history of Earth's crust-mantle system inferred from hafnium isotopes in chondrites. Nature 4421:931–933

Byerly GR, Lowe DR (1994) Spinels from Archaean impact spherules. Geochim et Cosmochim Acta 58:3469–3486

Czamanske GK, Himmelberg GR, Goff FE (1976) Zoned Cr-Fe spinels from the La Perouse layered gabbro Fairweather range Alaska. Earth Planet Res Lett 33:111–118

Farley KA, Montanari A, Hoemaker EM, Shoemaker C (1998) Geochemical evidence for a comet shower in the Late Eocene. Science 280:1250–1253

Glikson AY (1983) Geochemical isotopic and palaeomagnetic tests of early sial-sima patterns: the Precambrian crustal enigma revisited. Geol Soc Am Mem 4:183–219

Glikson AY (2007) Early Archaean asteroid impacts on Earth: stratigraphic and isotopic age correlations and possible geodynamic consequences In: Van Kranendonk MJ, Smithies H, Bennett VC (eds) Earth's oldest rocks. Developments in Precambrian Geology 15

Glikson AY, Allen C (2004) Iridium anomalies and fractionated siderophile element patterns in impact ejecta, Brockman iron formation, Hamersley Basin, Western Australia: evidence for a major asteroid impact in *simatic* crustal regions of the early Proterozoic earth. Earth Planet Sci Lett 20:247–264

Koeberl C, Reimold WU, Boer RH (2003) Geochemistry and mineralogy of early Archaean spherule beds, Barberton Mountain Land, South Africa: evidence for origin by impact doubtful. Earth Planet Sci Lett 119:441–452

Koeberl C, Reimold WU, Shirley SB (1996) Re-Os isotope and geochemical study of the Vredefort Granophyre: clues to the origin of the Vredefort structure South Africa. Geology 24:913–916

Kyte FT, Smit J (1986) Regional variations in spinel compositions: an important key to the Cretaceous/Tertiary event. Geology 14:485–487

Kyte FT, Shukolyukov A, Lugmair GW, Lowe DR, Byerly GR (2003) Early Archaean spherule beds: chromium isotopes confirm origin through multiple impacts of projectiles of carbonaceous chondrite type. Geology 31:283–286

Kyte FT, Zhou L, Lowe DR (1992) Noble metal abundances in an early Archaean impact deposit. Geochim Cosmochim Acta 56:1365–1372

Lugmair GW, Shukolyukov A (1998) Early solar system timescales according to 53Mn-53Cr systematics. Geochim et Cosmochim Acta 62:2863–2886

Macdonough WF, Sun SS (1995) Composition of the Earth. Chem Geol 20:223–253

Margolis SV, Claeys PF, Kyte ET (1991) Microtektites, microkrystites and spinels from a late Pliocene asteroid impact in the Southern Ocean. Science 251:1S94–1597

McCulloch MT, Bennett VC (1994) Progressive growth of the Earth's continental crust and depleted mantle: geochemical constraints. Geochim et Cosmochim Acta 58:4717–4738

Melosh HJ, Vickery AM (1991) Melt droplet formation in energetic impact events. Nature 350:494–497

Miura Y, Uyedo Y, Kedves M (2001) Formation of Fe –Ni particles by impact process. Proc Lunar Planet Sci Conf XXXII Abstract 2149

Montanari A, Koeberl C (1993) The Late Eocene Earth: Hothouse, Icehouse, and Impacts. Geol Soc Am spec Pap 452:322

Reimold WU, Koeberl C, Johnson S, Mcdonald I (2000) Early Archaean spherule beds in the Barberton Mountain land South Africa: impact or terrestrial origin? In: Koeberl C, Gilmour I (eds) Impacts and the early Earth. Springer, Berlin, pp 100–116

Robin E, Boclet D, Bonte P, Froget L, Jehanno C, Rocchia R (1991) The stratigraphic distribution of Ni-rich spinels in Cretaceous-Tertiary boundary rocks at El-Kef (Tunisia) Caravaca, Spain, and Hole-761C (Leg-122). Earth Planet Sci Lett 107:715–721

Scheffers A, Kelleat D (2002) Sedimentologic and geomorphologic tsunami imprints worldwide—a review. Earth Sci Rev 63:83–92

Schoenberg R, Kamber B, Collerson KD, Moorbath S (2002) Tungsten isotope evidence from ~ 3.8 Gyr metamorphosed sediments for early meteorite bombardment of the Earth. Nature 418:403–405

Shukolyukov A, Lugmair G W (1998) Isotopic evidence for the Cretaceous—tertiary impactor and its type. Science 282:927–929

Shukolyukov A, Kyte FT, Lugmair GW, Lowe DR, Byerly GR (2000) The oldest impact deposits on Earth. In: Koeberl C, Gilmour I (eds) Lecture notes in Earth science 92: impacts and the early Earth, Springer, Berlin, pp 99–116

Simonson BM, Davies D, Wallace M, Reeves S, Hassler SW (1998) Iridium anomaly but no shocked quartz from late Archaean microkrystite layer: oceanic impact ejecta? Geology 26:195–198

Sleep N, Zahnle KJ, Kasting JF, Morowitz HJ (1989) Annihilation of ecosystems by large asteroid impacts on the early Earth. Nature 342:139–142

Tredoux M, DeWitt MJ, Hart RJ, Armstrong RA, Lindsay NM, Sellchop JPF (1989) Platinum group elements in a 3.5 Ga nickel—iron occurrence: possible evidence of a deep mantle origin. J Geophys Res 94:795–813

Turekian KK (1982) Potential of 1870/1860s as a cosmic versus terrestrial indicator in high iridium layers of sedimentary strata. In: Silver LT, Schultz PH (eds) Geological implications of impacts of large asteroids and comets on the Earth. Geol Soc Am Sp Pap 190:243–249

Wallace M, Gostin VA, Keays RR (1990) Acraman impact ejecta and host shales—evidence for low-temperature mobilization of iridium and other platinoids. Geology 18:132–135

Zahnle K, Grinspoon D (1990) Comet dust as a source of amino acids at the Cretaceous/Tertiary boundary. Nature 348:157–160

Zhao M, Bada JL (1989) Extraterrestrial amino acids in Cretaceous/Tertiary boundary sediments at Stevns Klint, Denmark. Nature 339:463–465

Chapter 8
Precambrian Asteroid Impacts

Abstract Further to the few very large Precambrian impact structures identified (Maniitsoq, Yarrabubba, Vredefort, Sudbury), the application of geochemical, mineralogical and isotopic criteria to the study of impact ejecta allows documentation of large part of the early terrestrial impact record, yet it is suggested the known impacts constitute the 'tip of the iceberg' relative to the complete impact record.

The late heavy bombardment (LHB) of the Moon at about ~3.85 – 3.95 Ga (Ryder 1990, 1991,1997), evidenced by the large Mare basins (*Imbrium, Tranquilitatis, Serenitatis, Crisium, Orientalis*) has more than likely involved the Earth, with which the Moon was coupled during the first hundred million years since accretion from the solar nebula (Ringwood 1986). However, far from signifying the end of the Late Heavy Bombardment, the discovery of at least 9 major layers of microkrystite spherule-bearing ejecta ranging from 3,482 to 3,225 Ma in the Barberton greenstone belt and ~3.47 Ga equivalents in the Pilbara Craton, Western Australia (Lowe and Byerly 2010), likely represents ongoing bombardment in the Earth-Moon system. In total of near to 20 impactite units and impact structures are known from the Archaean and early Proterozoic, although some of the ejecta units may correlate (Table 1.2). Discoveries of impact ejecta were allowed by observation of millimeter-scale spherules in Archaean sediments in South Africa and Western Australia, identical with microkrystite spherules as defined by Glass and Burns (1988) from the 65 Ma Cretaceous–Tertiary asteroid impact boundary (Lowe and Byerly 1986, 1990; Lowe et al. 1989, 2003; Kyte et al. 1992; Byerly and Lowe 1994; Shukolyukov et al. 2000). Whereas field identification of microkrystites is hampered by their millimeter-scale, the common occurrence of tsunami deposits above the spherule units provides a helpful criterion (Simonson and Hassler 1997; Glikson 2004). Simonson and colleagues (Simonson 1992; Simonson and Hassler 1997; Simonson et al. 2000, 2009, 2010; Simonson and Glass 2004; Hassler et al. 2011) identified several impact spherule units and associated tsunami deposits at four stratigraphic levels (~2.63, ~2.56, 2.54, 2.48 Ga) in the Hamersley Basin of Western Australia and the Transvaal Basin, South Africa (Fig. 10.2).

Estimates of bolide size based on layer thicknesses, spherule diameters and the Ir and $\varepsilon^{53}Cr$ fluxes suggest asteroid diameters on the scale of tens of kilometer (Kyte et al. 1992; Lowe et al. 2003; Melosh and Vickery 1991). The craters from

A. Y. Glikson, *The Asteroid Impact Connection of Planetary Evolution*,
SpringerBriefs in Earth Sciences, DOI: 10.1007/978-94-007-6328-9_8,
© The Author(s) 2013

which ejecta has been derived have likely mostly been obliterated by metamorphism and anatexis or remain unknown. No signatures of impact have to date been found in the oldest Archaean terrains, including the Isua supracrustal sequence (~3.8 Ga) (Nutman et al. 2007) and the Akilia tonalite (~3.83 Ga) (Mojzsis and Harrison 2002). The deformed and amphibolite facies metamorphosed state of these sequences renders detection of impact ejecta difficult.

Major Archaean and early Proterozoic impact events documented by impact ejecta units and by impact craters were dated at ~3,482 Ma (BGB), ~3,472 Ma (2 units) (BGB and Pilbara), ~3,445 Ma (BGB), ~3,416 Ma (BGB), ~3,334 Ma (B GB), ~3,256 Ma, ~3,243 Ma, ~3,225 Ma (2 units) (BGB), ~2,975 Ma (Maniitsoq impact), ~2,647 Ma (Monteville) ~2,629 Ma (Jeerinah and Carawine), ~2,581 Ma (Reivilo), ~25.6 – 2.59 Ga (Paraburdoo), ~2.56 – 2.59 Ga, ~2.54 Ga (SMB) (2 units), ~2,481 Ma (Dales Gorge) and (<2,516 Ma, Kuruman), <2.65 Ga (Yarrabubba impact), ~2023 Ma (Vredefort impact), 2.13 – 1.85 Ga (Granesco, Greenland) and ~1,850 Ma (Sudbury impact) (Sect. 9.3; Table 1.1).

8.1 Archaean Impact Ejecta/Fallout Units

8.1.1 Early-Mid Archaean (~3.48–3.34 Ga) Impacts

Newly discovered mid-Archaean impact layers (Lowe and Byerly 2010) include 3482 – 3225 Ma-old microkrystite spherule-bearing units identified in the Barberton greenstone belt (BGB), Kaapvaal Craton, Transvaal (Fig. 3.3). The BGB ~3,472 Ma spherule unit correlates with a microkrystite spherules-bearing unit dated as 3467 ± 4 Ma within the ~3,474 – 3463 Ma Mount Ada Basalt in the central Pilbara Craton (Figs. 3.2, 3.3, 8.1). The Mount Ada Basalt comprises carbonated pillowed tholeiitic, high-Mg basalt, peridotitic komatiites, dolerite sills, minor volcaniclastic rocks, and intercalated sediments (Glikson and Hickman 1981; Hickman 1981; Van Kranendonk 2000; Van Kranendonk et al. 2002). The sequence includes chert, jaspillite, arenite, chert diamictite, and intraclast conglomerate, which overlie felsic volcanics/hypabyssals correlated with the Duffer Formation in the Marble Bar area. Lowe and Byerly (1986) reported millimeter to sub-millimeter-scale silicified spherules containing quench-crystallization and glass-devitrification textures from the antarctic chert member (ACM) in the Miralga Creek (Fig. 8.2). The spherules are distributed as streaks and lenses in chert and arenite over a north–south strike distance of at least 2.5 km, interrupted by numerous small faults (Glikson et al. 2004). Byerly et al. (2002) report a $^{207}Pb/^{206}Pb$ age of 3470.1 ± 1.9 Ma for euhedral zircons derived from the spherule-bearing unit, which are stratigraphically correlated with sediments overlying the Duffer Formation. A similar unit occurs in the Hoogenoeg Formation, Barberton greenstone belt, yielding a $^{207}Pb/^{206}Pb$ zircon age of 3470.4 ± 2.3 Ma. An impact origin of the spherules is corroborated by high sphericities, size distribution, quench textures, and inward-radiating K-feldspar mantles. These textures

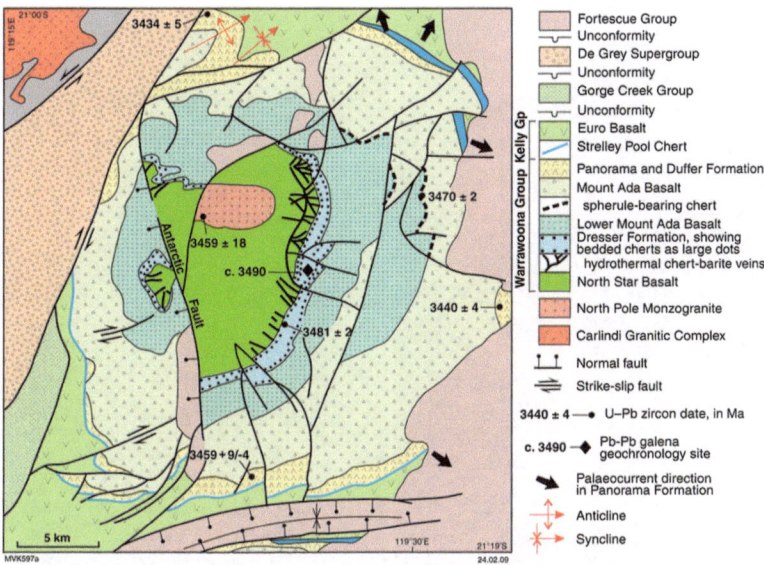

Fig. 8.1 Geological sketch map of the North Pole dome, central Pilbara Craton, Western Australia, showing the distribution of the ~3.47 Ga impactite unit in the middle of the Mount Ada Basalt sequence in the northeastern part of the dome (courtesy M. Van Kranendonk, Geological Survey of Western Australia, by permission)

are distinct from those of coexisting porphyritic, microlitic, and quench-textured angular volcanic fragments. High Ni, Cr, Ni/Co and Ni/Cr ratios of microkrystite spherules relative to Archaean tholeiitic and high-Mg basalts (~1.7–3.0) may serve as effective discriminants between these materials (Glikson et al. 2004). About 250 m stratigraphically below Lowe and Byerly's (1986) type locality, an interval occupied by hypabyssal intrusives, is a 0.6–0.8 m-thick unit of silicified breccia-conglomerate and chert-intraclast conglomerate (Fig. 8.2). The unit includes <40 cm-large subangular to sub-rounded chert pebble- to boulder-size fragments set in a sand to granule-size matrix, which includes sparse, but well-preserved, silicified and matrix resorbed microkrystite spherules. Pebbles and cobbles are dominated by chert-layer fragments, are poorly sorted, and show little imbrication. The arenite shows little or no current bedding. Since deposition in a shallow-water environment can be expected to have resulted in winnowing, corrosion, and destruction of the originally glassy spherules, their excellent preservation suggests deposition below wave-base level, likely reflecting a deep-amplitude tsunami event.

This evidence, supported by the pillowed subaqueous nature of thick basaltic sequences of the Mount Ada Basalt, testifies to a rapidly subsiding basin where subaqueous conditions were maintained. This contrasts with the shallow-water evidence provided by the stromatolite-bearing arenite-barite association of the Dresser Formation, located stratigraphically about 3,000 m below the ACM, or the stromatolite-bearing Strelley Chert, which overlies felsic volcanics of the Panorama

Fig. 8.2 ~3.47 Ga impact ejecta units, Antarctic Chert Member, Mount Ada Basalt, Miralga Creek, North Pole dome, Pilbara Craton. **a** Type locality of the impactite (dolerite (*DOL*), Antarctic Chert Member (*ACM*), spherule unit 1 (*S1*), spherule unit 3 (*S3*), felsic volcanics (*FV*), ferruginous chert/jaspilite (*FeCh*). **b** Jaspilite at the top of the ACM overlying the impactite. **c** Microkrystite spherule-bearing chert diamictite (spherules (*sp*), chert fragment (*Ch fr*), black chert fragment (*Bc.fr*). **d** Microkrystite spherule-bearing breccia-conglomerate (AJES, Geological Society of Australia, by permission)

Formation higher up in the sequence (Lowe 1980; Dunlop and Buick 1981). The accumulation of a ~6,000 m-thick sequence of mafic–ultramafic pillowed volcanics and associated hypabyssals suggests rapid subsidence tectonics associated with massive volcanism interrupted by impacts. In contrast to the preservation of impactites in the subaqueous below-wave base environment, evidence for impact fallout in shallow-water environments would have been removed through current reworking/winnowing of the delicate spherules. This environmental constraint, coupled with the difficulty in identifying the sub-millimeter spherules in the field, suggests that the impact record identified to date constitutes a minimum estimate.

8.1.2 A Mid-Archaean (~3.26–3.24 Ga) Impact Cluster and Related Tectonic and Igneous Events

The identification of a 3256 – 3225 Ma asteroid impact cluster associated with the sharp break between the ~3.482 – 3.334 Ga mafic–ultramafic volcanic sequence

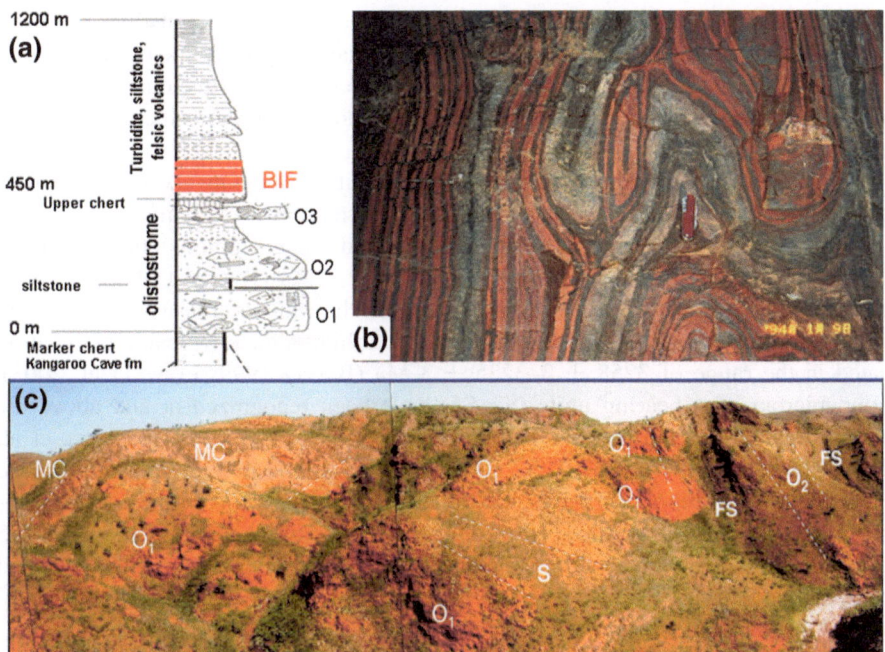

Fig. 8.3 Pilbara correlations with the Barberton ~3.24 Ga impactites level. **a** Schematic columnar section of the transition from volcanics of the Sulphur Springs Group (dacite of Kangaroo
Caves Formation: 3235 ± 3 Ma), capped by Marker Chert, overlain by three olistostrome units
separated by ferruginous shale (olistostrome units 1–3 ($O_{1–3}$). **b** Jaspilite of the Paddy Market
Formation, Copin Crossing. **c** Large blocks of chert, siltstone and siliceous volcanics in the
Sulphur Springs area, central Pilbara Craton, occupying a stratigraphic position similar to that
of ~3.24 Ga impact fallout units in the Barberton Mountain land, South Africa, signifying major
faulting near-contemporaneous with asteroid impacts (olistostrome units 1–3 ($O_{1–3}$), shale (*S*),
ferruginous shale (*FS*), Marker Chert (*MC*)

of the Onverwacht Group and the <3225 Ma turbidite-banded iron formation–
felsic volcanic Fig Tree Group, Barberton Greenstone Belt (Fig. 3.3) (Lowe and
Byerly 1986; Lowe et al. 1989, 2003; Kyte et al. 1992; Byerly and Lowe 1994;
Shukolyukov et al. 2000; Glikson 2007b) herald a new insight into the origin of
Archaean greenstone belts. This well-documented asteroid impact cluster and
correlated units in the Pilbara Craton (Figs. 8.2, 8.3) (Glikson and Vickers 2005;
Glikson 2008) provide evidence for tectonic and magmatic events and major
unconformities triggered by very large asteroid impacts.

The Barberton spherules possess a spectrum of features diagnostic of microkrystites, including Inward-radiating quench textures, centrally offset vesicles
reflecting aerodynamic effects, quench-textured and octahedral Ni-chromites
(NiO < 23 %) with high Co, Zn and V abundances, unknown in terrestrial chromites (Byerly and Lowe 1994), and chromite-hosted PGE nano-nuggets (Kyte
et al. 2003). PGE chondrite-normalized patterns display marked depletion in

volatile species (Pd, Au) relative to refractory species (Ir, Pt), as distinct from ter-restrial PGE profiles, excepting depleted mantle harzburgites (Glikson 2005) (Fig. 6.2). Negative $^{53}Cr/^{52}Cr$ isotopic indices ($\varepsilon^{53}Cr = -0.32$) correspond to values of carbonaceous chondrites and K–T boundary impact fallout deposits, distinct from terrestrial values.

The uppermost mafic–ultramafic volcanic unit of the Onverwacht Group is represented by the Mendon Formation, an assemblage of komatiitic volcan-ics and hypabyssal and altered equivalents capped by ferruginous chert, dated by U–Pb zircon from a middle chert unit as 3298 ± 3 Ma (Byerly 1999) (Fig. 3.4). Unconformably above the Mendon Formation is the Mapepe Formation, the basal unit of the Fig Tree Group which comprises a turbidite–felsic volcanic association dated in the range of 3258 ± 3–3225 ± 3 Ma (Byerly 1999; Lowe et al. 2003). Four microkrystite-bearing units (S2–S5) have been recognized at and above the unconformable base of the Fig Tree Group (Lowe and Byerly 1986, Lowe et al. 2003; Lowe and Byerly 2010). S2 coincides with the base of the clastic Mapepe Formation (Fig Tree Group) whereas S3–S5 spherules occur within clastic sedi-ments and felsic pyroclastic 110–120 m above the basal contact. In places S5 occurs unconformably above deeply incised Mendon Formation komatiites, injected by spherule-bearing breccia dykes and black chert veins to a depth of 100 m, repre-senting seismic and tsunami effects (Lowe and Byerly 1999; Lowe et al. 2003).

The juxtaposition of impact spherule units with the boundary between the mafic–ultramafic Onverwacht Group and the semi-continental sequence of the Fig Tree Group suggests regional and possibly global tectonic reorganization associ-ated with the ~3.2 impact cluster. Evidence consistent with the suggestion of Lowe et al. (2003) is provided by U–Pb zircon ages of the Nelshoogte trondhjemite (3236 ± 1 Ma) and the Kaap Valley tonalite (3227 ± 1 Ma), which closely fol-low S3–S5 impact units (3243 ± 4 Ma). The high precision of these ages suggests that anatexis and plutonic ascent and crystallization of granitoid magmas occurred within a period of ~7–16 m.y. following the impacts, represent lag plutonic activ-ity. The tonalite–trondhjemite geochemistry of the granitoids suggests anatexis of mafic crustal sources (Glikson 1979, 1984).

To date With the exception of the Pilbara ~3.47 Ga ACM spherules no impact ejecta units contemporaneous with those of the Barberton Greenstone Belt have been encountered in other cratons. However, studies of the boundary between the ~3,255 – 3,235 Ga volcanics of the Sulphur Springs Group and the turbidite-banded iron formation–felsic volcanic assemblage of the Gorge Creek Group, Pilbara Craton, display isotopic age and stratigraphic correlations with the Barberton ~3.2 Ga-old impact boundary (Figs. 8.3, 8.4).

The Sulphur Springs Group (SSG) (Van Kranendonk et al. 2002; Vearncombe et al. 1998; Brauhart et al. 1998) is dominated by a volcanic succession uncon-formably overlying volcanics of the 3350 – 3310 Ma Euro Basalt. The Euro Basalt is overlain by conglomerate, wackes and felsic volcanic rocks, including spher-ulitic tuff (Leilira Formation, 3325 – 3255 Ma) (Buick et al. 2002). The Leilira Formation is overlain by komatiite (Kunagunarrina Formation), basalt–andesite, rhyolite and an uppermost unit of dacite–rhyolite (Kangaroo Caves Formation)

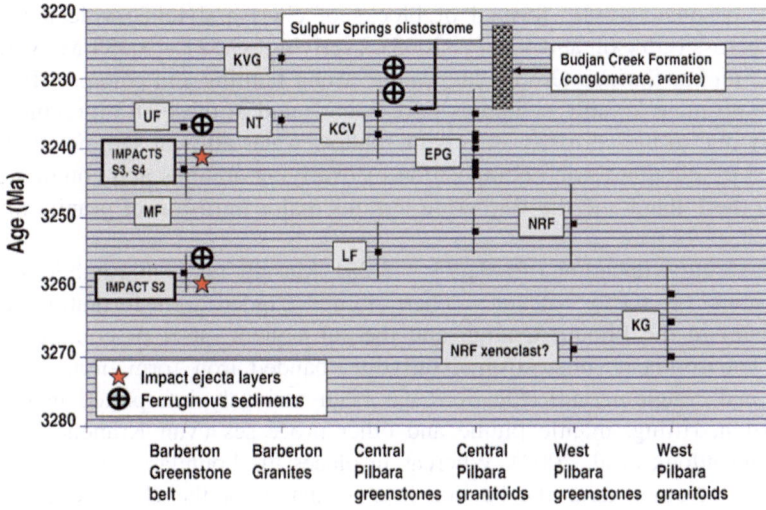

Fig. 8.4 Isotopic age correlations between ~3.28 and ~3.22 Ga units in the Kaapvaal Craton and Pilbara Craton. *Solid squares* and *error bar lines*—U–Pb ages of volcanic and plutonic units, *stars* impact ejecta layers, *circled crosses* ferruginous sediments, Mapepe Formation (*MF*), Ulundi Formation (*UF*), Nelshoogte tonalite (*NT*), Kaap Valley Granite (*KVG*), Leilira Formation (*LF*), Kangaroo Cave volcanics (*KCV*), Eastern Pilbara granites (*EPG*), Nickol River Formation (*NRF*) (including an older xenoclast), Karratha Granite (*KG*) (Elsevier, with permission)

(3251– 3235 Ma), capped by an up to 50 meters-thick chert horizon composed of silicified fine-grained epiclastic and siliciclastic rocks. The chert formed during hydrothermal circulation associated with the final emplacement of the syn-volcanic Strelley Granite laccolith and precipitation of volcanogenic massive sulphide deposits (Vearncombe et al. 1998; Brauhart et al. 1998). In the Sulphur Springs area the chert is overlain by a large- block mega-breccia, siltstone, turbidites, fine-grained clastics and banded iron formation correlated with the Nimingarra Iron Formation (Williams 2003) of the Pincunah Hill Formation, Gorge Creek Group. The sequence includes felsic volcanic breccia consisting of spherulitic volcanic rocks. In the Sulphur Springs area the Pincunah Hill Formation varies laterally in thickness and pinches out at the apex of the syn-volcanic Strelley Granite. At this locality, the overlying Corboy Formation fills in a submarine canyon as a series of turbidites that lap onto the marker chert, which had previously been lithified, as evidenced by sandstone dykes filled by Corboy sandstones in the marker chert. Turbiditic feldspathic arenite beds grading up to argillite of the Corboy Formation vary in thickness across faults, suggesting horst and graben fault activity during sedimentation, with implications for a strongly dissected relief.

The search for equivalent 3.256–3.225 Ga microkrystite spherule units in the Pilbara Craton indicates that correlated stratigraphic units occur at (1) above an unconformity where Leillira Formation conglomerate and arenite (3325 – 3310 Ma) overlies Euro Basalt (3346 – 3335 Ma) and (2) at a large boulder deposit, or olistostrome, formed of blocks of chert and felsic volcanics up to 250 m across (Glikson and Vickers 2005), overlying ~3,521 – 3,235 Ma volcanics off the Sulphur Springs Group, suggesting strong faulting and collapse (Fig. 8.3). To date no microkrystite spherules have been found at these stratigraphic levels, possibly due to the corrosive effects of shallow water alluvial sedimentation represented by the quartz-rich composition of overlying arenites and conglomerates. In both the Pilbara and the Barberton terrains major intrusion of granite magmas occurs at ~3.24 Ga.

The abrupt truncation of ~3,275 – 3,251 Ma-old mafic–ultramafic volcanic successions (+ felsic volcanics, chert, minor banded iron formation, arenite and shale) in the Pilbara Craton, and the subsequent onset of continental-type crustal environment (olistostrome, turbidite, banded iron formation, felsic volcanics and conglomerate) (Fig. 8.3) has been variously interpreted in terms of subduction, rifting, mantle plume and other processes (Van Kranendonk 2000; Van Kranendonk et al. 2002). Whereas in places the boundary appears paraconformable, in other areas the banded iron formation of the Gorge Creek Group (Nimingarra Iron Formation) overlies basal conglomerate above granitoid of the Muccan Batholith, dated as 3255 – 3244 Ma.

Based on stratigraphic and isotopic age the following correlations pertain (Figs. 3.3, 8.4):

a. Between volcanics of the Mendon Formation (BGB) (3334 – 3256 Ma) and the overlying BGB S2 spherule unit (~3,243 Ma) and the Sulphur Springs mafic–ultramafic volcanic sequence (3273–3255 Ma)
b. Between the S3–S5 spherule units of the lower part of the Fig Tree Group (Mapepe Formation: 3256 – 3225 Ma) and the 3235 ± 3 Ma age (Buick et al. 2002) of the volcanic Sulphur Springs Group, defining the oldest age limits of the sedimentary-tuff sequence of the Gorge Creek Group (Figs. 8.3, 8.4).

Major plutonic igneous activity, producing a ~3.25 – 3.235 Ga suite of granitoids defined as the Cleland Supersuite is documented throughout the Pilbara including central parts of the Mount Edgar, Muccan and Warrawagine batholiths (east Pilbara), Carlindi batholith, Strelley Granite and Tambourah Monzogranite (central Pilbara). Similar associations are observed in the Roebourne Group, west Pilbara Craton, where komatiite and basalt of the ~3,270 Ma Ruth Well Formation are overlain by 3270 – 3250 Ma sandstone, shale, banded iron formation, and felsic volcanics of the Nickol River Formation.

Hints the ~3.2 Ga impact cluster may have affected the Moon may be furnished by isotopic Ar–Ar dating of lunar spherules (Culler et al. 2000) which, when combined with earlier Rb–Sr and Ar–Ar isotopic studies of lunar basalts (BVTP 1981), suggests the occurrence of near-contemporaneous volcanic activity such as in Oceanus Procellarum (3.29 – 3.08 Ga) and the Hadley Apennines (3.37 – 3.21 Ga) (Glikson 2001). Such correlations are consistent with laser $^{40}Ar/^{39}Ar$ analyses of lunar impact spherules showing a significant age spike at 3.18 Ga near the boundary of the Late Imbrian (3.9 – 3.2 Ga) and the Eratosthenian (3.2 – 1.2 Ga). With very poor error margins 34 lunar impact spherules yield a mean age of 3188 ± 198 Ma with a median age at 3181 Ma. 7 spherule ages yield a mean age of 3178 ± 80 Ma and a median at 3186 Ma. Combined evidence suggests the period 3.24 ± 0.1 Ga experienced a major impact cataclysm in the Earth–Moon system leading to renewed volcanic activity in some of the lunar Mare basins. Large errors in the Ar–Ar and Rb–Sr isotopic ages preclude precise correlations with terrestrial events.

8.1.3 The Late Archaean (~2.63–2.48 Ga) Impact Cluster

8.1.3.1 The 2.63 Ga Jeerinah Impact Layer and Carawine mega-breccia

The Jeerinah Impact Layer (JIL), located toward the top of a sequence of silt-stone, chert, and mafic volcanics termed the Jeerinah Formation, comprises a basal microkrystite-bearing rip-up siltstone fragment breccia, a lenticular spherule unit up to about 60 cm thick, and an overlying boulder-size debris flow conglomerate (Fig. 8.5b and 1.5a). The sequence represents an initial seismic and current disturbance of the seabed, settling of microkrystite spherules and subsequent slumps and debris flow. The original observation of a 6 mm-thick lamina of microkrystite spherules above black shale of the Jeerinah Formation at Ilbiana Well (Simonson et al. 2000), was followed by its discovery at Hesta Siding on the Newman-Port Hedland railway line, about 50 km to the northwest of Ilbiana Well (Simonson et al. 2001). The age of the JIL is constrained by U–Pb zircon date of 2629 ± 5 Ma of overlying volcanic tuff (Nelson et al. 1999) and U–Pb ages on zircon from the base of the Jeerinah Formation (2684 ± 6 Ma, 2690 ± 6 Ma; Arndt et al. 1991). The Hesta Siding section displays a transition from laminated argillite of the Jeerinah Formation upwards into a ~5 m-thick argillite-chert unit, which underlies a ~1.7 m-thick impact fallout unit (Fig. 8.5). The latter consists of:

a. A lower ~40 cm-thick zone of microkrystite spherule-bearing siltstone breccia consisting of a poorly sorted assemblage of rip-up red siltstone fragments set in a microkrystite spherule-rich matrix (Fig. 1.5a). The fragments are identical in composition to the underlying chert-argillite sequence of the Jeerinah Formation.

b. A ~60 cm-thick zone dominated by microkrystite spherules (Fig. 1.5a, b) mixed with argillite and chert intraclasts.

c. A ~70 cm-thick boulder breccia dominated by angular to rounded cobbles and boulders of chert reaching ~50 cm in size (Fig. 8.5c). The breccia is capped by a ~60 cm-thick argillite overlain by weathered ironstones of the Marra Mamba Iron Formation.

The sequence signifies impact triggered seismically induced disruption of the seabed involving excavation of rip-up clasts concomitant with settling of micro-krystite spherules and microtektites. The waning of impact fallout was succeeded by debris flow consisting of up to boulder-size angular to sub-rounded chert

Fig. 8.5 Jeerinah impact layer (*JIL*), Hesta type section, central Pilbara Craton. **a** Columnar section of the Jeerinah impact layer (*JIL*) and host sediments.**b** Railroad-side exposure at Hesta siding, central Pilbara, showing Jeerinah formation siltstone (*JS*) and siltstone–chert (*JSC*), the Jeerinah impact layer (*JIL*) and laterite-altered base of the marra mamba iron formation (*MM*). **c** The Jeerinah impact layer overlain by mega-breccia. **d** Boulder debris flow above JIL, containing tabular and rounded decimeter-scale fragments. Swiss knife is 8 cm. Feldspar-dominated micro-tektites (large fragment at the *bottom*)

fragments, followed by a return to deposition of argillite, tuff, and ironstones. The lack of strong current signatures in the underlying shale and siltstones of the Jeerinah Formation and in the overlying ironstones suggests deposition of JIL was related to an abrupt high-energy seismic and tsunami pulse affecting an environment otherwise located below wave base. Whereas the basal rip-up clasts represent initial seismic effects, the debris flow that overlies the spherule unit is related to a deep amplitude tsunami wave triggered by the impact.

In the eastern Pilbara Craton a stratigraphic-consistent ~10–30 meters-thick microkrystite-bearing mega-breccia unit composed of fragments and boulders of dolomite and chert occurs near the base of the Carawine Dolomite (Simonson 1992; Hassler et al. 2000; Williams 2003; Glikson et al. 2004). The strata-bound mega-breccia unit extends over a strike distance of nearly 100 km northwest-southeast between Ripon Hills and Warrie–Warrie Creek areas in the eastern outlier of the Hamersley Basin (Figs. 8.6, 8.7). It occurs within the lower basin facies of the Carawine Dolomite, about 30–100 m above the contact with underlying siltstone of the Jeerinah Formation and is either excavated into, or

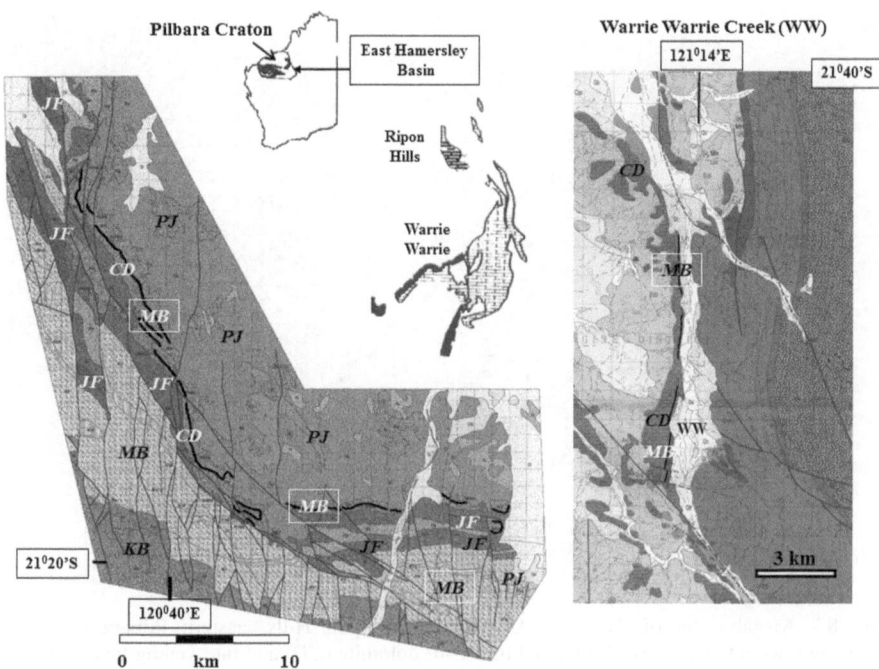

Fig. 8.6 Geological sketch maps of the Eastern Hamersley Basin, indicating the distribution of the megabreccia (*MB*) of the carawine dolomite (*CD*). Kylena Basalt (*KB*), Maddina Basalt (*MB*), Jeerinah Formation (*JF*) ~2.629 ± 5 Ga), Pinjian Chert (*PJ*) (silicified surface breccia), Gregory Range (*GR*), Warrie Warrie (*WW*) Creek belt, Ripon Hills (*RH*). The MB unit is shown as a thick black line. Modified after Williams (2003) (Astrobiology, by permission)

conformably overlies, layered carbonate. The top of the mega-breccia pile is invariably conformable and commonly includes lenses of microkrystite spherules and microtektites.

The mega-breccia consists of chaotically structured randomly oriented angular fragments, blocks, and bed segments of dolomite on scales ranging up to several meters large (Fig. 8.7). Blocks may be oriented at high angle to the overall boundary of the mega-breccia or may show a crude subparallel imbrication of fragments. Excavation of the mega-breccia involved dislodgment of up to many meters-large bedding segments of dolomite and chert, and less commonly of siltstone, which are separated and injected by microbreccia breccia veins, wedges, and layer-parallel sheets. Ductile deformation of bed segments is observed on a range of scales, for example, in layered chert-carbonate blocks and as small folded concretions, suggesting these blocks were only semi-consolidated when the tsunami hit.

Hassler et al. (2000) identified microtektite-bearing matrix within the mega-breccia where, remarkably, microkrystites and microtektites are preserved intact in injected breccia veins despite expected friction-induced corrosion, likely due to hydraulic dilation of the injected veins under extreme tsunami pressures. Features

Fig. 8.7 Megabreccia of Carawine Dolomite at Rippon Hills, eastern Pilbara Craton. **a** Megabreccia containing large horizontal blocks of dolomite (*CD*) and intervening layers of breccia (boundaries marked by *arrows*) connected by breccia veins (*V*). Swiss knife is 8 cm; **b** Large block of Carawine megabreccia (*CB*) underlain and overlain by breccia (*BR*) (AJES, Geological Society of Australia, by permission)

relevant to the interpretation of the origin of the mega-breccia include: (1) concentration of microkrystite spherules toward the top of the breccia pile indicates settling mostly postdated brecciation; (2) The chaotic structure of the mega-breccia, including meter- to many meter-scale blocks that rest conformably on, or as depressions within, the underlying layered carbonates, underpins the essentially autochthonous to semi-autochthonous nature of the mega-breccia; (3) The disruption of up to 7-meters-long bedding segments indicates high-energy tsunami-induced excavation of the sea floor; (4) Preservation of the megabreccia explained in terms of its location below wave base.

Hassler et al. (2000) and Hassler and Simonson (2001) developed a tsunami-generated model that includes the following stages (a) Settling of Microkrystite spherule and microtektites accompanied by current reworking of the sea floor; (b) arrival of tsunami wave systems within hours from the impact, causing substrate erosion and reworking expressed by cross-layering and transport of meter-scale rip-up clasts; (c) syn-tsunami to post-tsunami return flow and offshore transport of intra-basinal sediments and continental debris followed by reworking, resulting in hummocky cross stratification; and (d) down-slope gravity debris flow/slumping and turbidity currents, erosion, and redeposition of preexisting sediments. The lack of shock metamorphic features in detrital quartz associated with the mega-breccia and the presence of relic altered ferromagnesian elements, mainly chlorite, suggests impact mostly affected mafic target crust (Simonson et al. 1998). Tsunami propagation estimates by Hassler et al. (2000), assuming a 5 km-large projectile and ~60 km-diameter crater, was interpreted in terms of an oceanic impact located between 2,000 and 8,000 km from the Hamersley Basin. Tsunami wave/s could also have originated at impact-activated faults and/or plate boundaries. Estimates of crater location based on the time relations between spherule settling and tsunami arrival involve a number of assumptions, including the point origin of the tsunami and velocity of airborne spherules.

Submarine impact-induced disruption triggered by seismic shock and tsunami waves has been documented in the vicinity of the Chicxulub impact structure at Belize (Pope et al. 1997). An autochtonous to sub-autochtonous dislodgement of angular boulders is supported as allochtonous transport of boulders and fragments would have resulted in partly rounded blocks. In many outcrops angular blocks and large detached bedding segments of mega-breccia rest on, or are torn from, underlying excavated carbonates. The concentration of spherule lenses above the mega-breccia requires tsunami arrival has predated spherule settling from the ejecta cloud. However, the injection of spherules in veins into the breccia pile suggests settling from the ejecta cloud took place while disturbance and dilation pressures of the sea floor continued. Assuming deep ocean tsunami travel times in the order of 1,000 km/h and a relative proximity of the parent impact crater, the possibility of multiple tsunami waves cannot be discounted.

8.1.3.2 The ~2.57–2.56 Ga Impact Cluster

The Wittenoom Formation, Hamersley Basin, northwestern Australia, contains at least 3 impact fallout units, including the ~2.57 Ga Paraburdoo spherule layer (PSL) (Hassler et al. 2011), correlated with the Reivilo spherule layer (RSL) in the western Transvaal (Fig. 10.2), and the ~2.56 Ga spherule marker bed (SMB) (Simonson 1992; Simonson and Hassler 1997), which includes two spherule-tsunami sequences (Glikson 2004). The ~2.57 Ga spherule unit occurs within the carbonate-dominated Paraburdoo Member of the Wittenoom Formation and the SMB within the 230 m-thick siltstone-carbonate sequence of the Bee Gorge Member of the Wittenoom Formation. (Figure 8.8). The SMB forms a prominent unit of calcareous arenite up to 1.6 m thick featuring dark weathering surfaces and consisting of densely packed mm-scale spherules which form cm to decimeter-scale discontinuous lenses overlain by decimeter-thick turbidite units. The thickness of spherule unit generally decreases through the Hamersley Basin from north to south (Simonson 1992). The turbidities consist of argillaceous carbonate, carbonate-rich argillite, and immature arenite hosting silicate and carbonate fragments and varying proportions of microkrystite spherules ranging from ~95 % spherules to isolated spherules in argillite and arenite matrix. Spherule shells display inward-radiating K-feldspar fans (Fig. 8.8b), which may also occur at spherule centers, otherwise occupied by coarse-grained quartz, sericite, chlorite, carbonate, and Fe-oxide. Some spherules are entirely occupied by randomly oriented K-feldspar fans and/or feldspar microlites.

Fig. 8.8 The ~2.56 Ga spherule marker bed (*SMB*), Hamersley Basin. **a** Bee Gorge section, showing the SMB horizon above the Bee Gorge carbonate-shale member (*BGM*) of the Wittenoom Formation, below the sylvia formation (*SF*) (ferruginous shale and chert), below the Mt McRae ferruginous shale and banded iron-formation (*MMS*). **b** Microkrystite spherules of the SMB. Q quench K-feldspar crystallites, *V* outline of vesicle, devitrification (*d*) features of radiating K-feldspar crystallites. **c** Munjina Gorge section of the SMB showing two impact cycles, SMB-1 and SMB-2, overlain and underlain by carbonate, siltstone, and chert (*CSC*). Each cycle includes a basal layer or series of lenses of microkrystite spherules (*MKZ*) overlain by rhythmic turbidites (seismic zone, SZ), overlain by a cross rippled tsunami zone (*TZ*). The two cycles are separated by a stratigraphically consistent layer of silicified black siltstone denoted as a 'Quiet Zone' (*QZ*) (AJES, Geological Society of Australia, by permission)

Microtektites form irregular fragments larger than microkrystites and some microkrystites displaying transitional characteristics with K-feldspar-dominated angular microtektite fragments. K-feldspar is extensively replaced by carbonate and quartz. The turbidites contain flat rip-up intraclasts of ferruginous argillite and carbonate lutite, commonly up to about 5 cm long, but in places much larger (Simonson 1992). The high length/thickness ratio of these fragments suggests derivation from the near-autochthonous underlying substratum, reflecting seismic and/or current disruption of the seabed. The highly ferruginous composition of many clasts may represent oxidizing conditions at the sea floor prior to the impact or, more likely, intense secondary alteration of the porous spherule units.

Sedimentological studies of 15 SMB localities suggest a slope or deep shelf environment persisting below and above the impact fallout layer (Simonson 1992;

Hassler et al. 2000; Hassler and Simonson 2001). The spherules occur in discontinuous cross-layered lenses and cross-ripples which form connected bedforms up to 150 meters-long. Paleo-current measurements of cross-ripples and climbing-ripples suggest prominence of southward-directed palaeocurrents (Hassler et al. 2000), an observation consistent with (1) general thinning of the spherules from northwest to south-east and (2) occurrence in SMB turbidites of basement derived detrital quartz grains that show mosaic/annealed texture.

Simonson (1992) documented splitting of the SMB at its westernmost occurrences into three parts: (1) a lower argillite containing spherule-bearing lenses many decimeters long and up to 25 mm thick; (2) a middle 7-cm-thick laminated calc-argillite; and (3) upper 60-cm-thick cross-layered dolomitic turbidite unit. This author regards the SMB as having been emplaced by several closely spaced sedimentary pulses (Simonson 1992). Later observations confirm the presence of distinct spherule horizons within the SMB, each overlain by turbidity current pulsation (Glikson 2004) (Fig. 8.8c). Thus in the Munjina Gorge and Wittenoom Gorge areas, located about 40 km apart, SMB-1 and SMB-2 impactite cycles are separated by 1–50 cm-thick little-disturbed siliceous siltstone. SMB-1, the stratigraphically lower cycle, contains at its base a centimeter-scale spherule-rich horizon, or discontinuous 5 cm-thick spherule rich lenses. In many areas spherules are missing. The spherule unit is overlain by graded-bedded cycle of arenite-turbidites, in turn overlain by cross-layered arenite or by siltstone capped by convolute climbing cross-bedding and eddie structures, signifying turbulence. The SMB-1 and SMB-2 cycles are separated by a consistent black siltstone layer, generally about 5–10 cm thick but reaching 20 cm in thickness, representing a lull in tsunami activity.

SMB-2, the stratigraphically upper cycle, includes a ~20 cm-thick densely packed spherule layer or lenses at its base, which commonly contains intercalations of black siltstone or isolated microkrystite-bearing carbonate bands with containing mm-scale siliceous peloids. The spherule unit is overlain by current-rippled argillite representing lower current intensity as compared with current structures within the lower SMB-1 cycle. SMB-2 can be discriminated from SMB-1 thanks to its lighter-color carbonate-dominated cross beds and intercalated black siltstone layers and lenses (Fig. 8.8c). It is unlikely that the SMB-2 spherules resulted from re-deposition of SMB-1 spherules since (1) such reworking can be expected to result in corrosion of the delicate spherule structures, as contradicted by their near-perfect preservation; (2) no excavation of the underlying black siltstone marker is observed. By analogy with the SMB, the Monteville spherule layer (MSL) in the western Transvaal (Simonson et al. 1999) contains two spherule layers separated by 1.8 m of shale and carbonate (Hassler and Simonson 2001, p. 13) (Fig. 10.2). The age of the MSL is uncertain, with estimates made in the 2.54 – 2.55 Ga range and at ~2.64 Ga (Simonson et al. 2001). The evidence is thus consistent with closely spaced impactite units representing a ~2.57-2.54 Ga impact cluster.

U–Pb zircon ages for crystal tuff unit located about 75 m below the SMB yielded 2603 ± 7 Ma and 2561 ± 8 Ma (Trendall et al. 1998), with the former age corresponding to older re-deposited zircon xenocrysts. Woodhead et al. (1998) measured a Pb–Pb carbonate whole rock age of 2541 ± 18/–15 Ma on the SMB, within error from the U–Pb ages. Pb–Pb isotopic analyses of carbonate from the Wittenoom Formation yielded ages of 2505 ± 37 Ma and 2346 ± 38 Ma, respectively, representing diagenetic overprinting (Jahn and Simonson 1995).

8.1.3.3 The ~2.48 Ga Dales Gorge Impact

LaBerge (1966) documented millimeter-scale spherulitic textures within chert fragment-bearing ferruginated tuff of the Dales Gorge Member Shale Macroband No. 4 (DGS4), Brockman Iron Formation, Hamersley Basin, Western Australia. The DGS4 Macroband can be traced at least 30 km in the Dales Gorge–Wittenoom Gorge area, where it forms a 1–2 m-thick soft ledge or cliff recess within cliff-forming banded ironstones. Trendall and Blockley (1970) referred to this unit as a spherule-bearing breccia. The unit contains textural, mineralogical, and trace element evidence of an impact fallout, identified by Simonson (1992), who documented microkrystite spherules featuring inward-radiating K-feldspar fans surrounding stilpnomelane-dominated interiors (Fig. 8.9e). The unit includes tsunami transported fragments and boulders of chert, banded iron formation and carbonate (Fig. 8.10). Trendall et al. (2004) determined an age of ~2.48 Ga for the DGS4. Glikson and Allen (2004) conducted SEM analyses of the spherules, identifying extraterrestrial geochemical and mineralogical characteristics, including Nickel-rich nano-nuggets.

Outcrops of DGS4 occur at Dales Gorge, Yampire Gorge and Wittenoom Gorge. The unit forms a soft to crumbling, black-weathered, laminated to non-laminated zone, which can be easily mistaken for tachylite-rich pyroclastics (Fig. 8.9c, d). The spherule-bearing unit forms a conformable stratiform zone toward the top of a tuff-siltstone-siderite carbonate sequence, varying in thickness from about 10–30 cm. Where observed, the unit conformably overlies ferruginous tuff. At Wittenoom Gorge the spherule-bearing unit contains abundant palimpsest fragments consisting of stilpnomelane and chert fragments, ranging in size up to meter-scale boulders of chert, banded iron formation (BIF) and carbonate. In hand specimens the black-aphanitic to clastic spherule-bearing rock consists of altered fragments of stilpnomelane and isolated to tightly packed spherules, marked by white K-feldspar shells which display inward-radiating crystal fans and centrally offset vesicles (Fig. 8.9e). Microscopic observations reveal a wide range of morphologies, including oblate, disc-shaped, and near dumbbell-like spherules (Glikson and Allen 2004). The meteoritic origin of the spherules is demonstrated by unique mineralogical and geochemical features, including (1) Ni-rich metal, oxide, sulfide, and arsenide submicron particles within K-feldspar spherule shells (Fig. 8.9f); (2) high Ni levels in Fe-oxide grains; (3) high Ni abundances, very high Ni/Co ratios, and high Ni/Cr ratios in stilpnomelane spherule interiors, and (4) low Pd/Ir, Pt/Ir, and Pd/Pt ratios. These patterns reflect loss of the volatile PGEs (Pd, Au) upon atmospheric condensation of impact-released vapor.

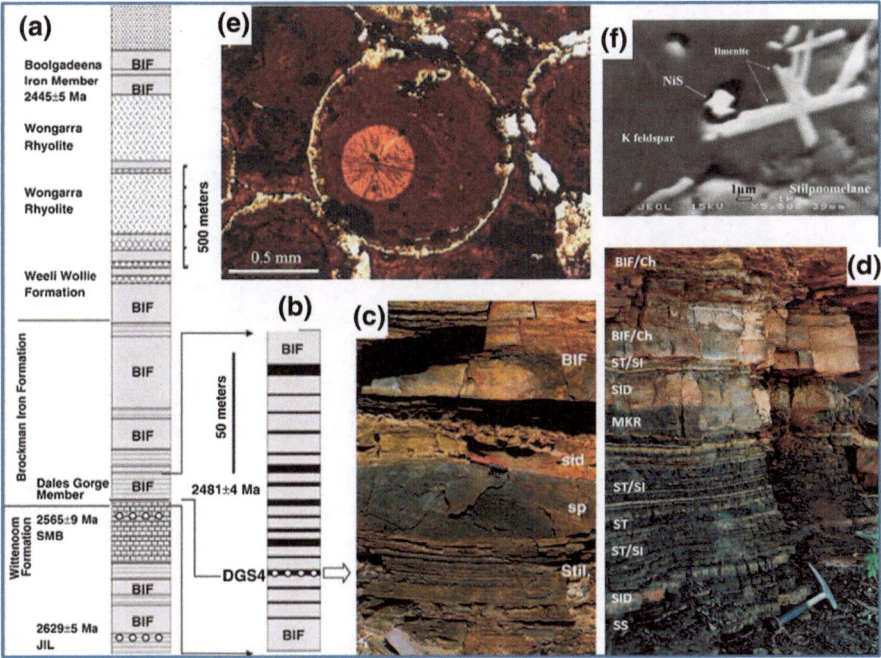

Fig. 8.9 - Impact ejecta of the ~2.48 Ga shale macroband of the Dales Gorge member, Brockman Iron Formation. **a** Schematic columnar diagrams showing main units of the Hamersley Group and U–Pb zircon isotopic ages (after Trendall et al. 2004). *JIL* Jeerinah Impact Layer, *SMB* Spherule Marker Bed. **b** Banded-iron macrobands and shale macrobands of the Dales Gorge Member; **c** Dale Gorge outcrop of impactite band (*sp*) at the top of DGS4, overlying finely laminated stilpnomelane-dominated tuff/sediment bands (*Stil*), and underlying siderite-rich siltstone band (*sid*) and banded iron-formation (*BIF*). **d** Yampire Gorge outcrop of impactite band (*MKR*) intercalated with banded iron formation (*BIF*), chert (*Ch*), stilpnomelane-rich layers (*ST*), siltstone (*SI*), siderite (*SID*) and sandstone (*SS*) layers. **e** Stilpnomelane dominated impact spherules (microkrystites) rimmed by thin light feldspar rims and cored by radiating centrally offset vesicle of stilpnomelane (AJES, Geological Society of Australia, by permission)

The isolated chert, BIF and carbonate fragments and boulders are interpreted in terms of exotic tsunami-transported blocks ripped off submarine scarps, possibly reflecting impact-triggered faulting. The mega-clasts appear to have been incorporated immediately following deposition of the spherule-rich material and only limited evidence is observed for imbricated debris flow. Likely equivalents of DGS4 were encountered in a drill hole ~30 km south of Griquastad, western Transvaal Basin, where a ~1 cm-thick spherule layer is located below an ~80 cm-thick breccia unit intercalated in a 2 meters-thick shale unit near the base of the Kuruman Iron Formation, about 37 meters above Gamahuann carbonates, relationships analogous to those of the DGS4.

Early Proterozoic impact ejecta and fallout units include ~2,130 – 1,848 Ga Palaeoproterozoic impact spherules in a dolomite layer in the Ketilidian orogen,

Fig. 8.10 A boulder of banded chert (*CB*) incorporated in the upper part of DGS4, resting in part on a spherule-bearing stilpnomelane matrix (*MKR*) containing chert fragments. Swiss knife is 8 cm

South Greenland (Chadwick et al. 2000) and impact ejecta related to the 1.85 Ga Sudbury impact structure (Addison et al. 2005; Jirsa et al. 2008; Cannon et al. 2010) (Sect. 9.3; Table 1.2).

8.1.4 Impact-Triggered Tsunami Events

Impactite bearing sedimentary units in the Pilbara Craton are associated with a spectrum of high-energy sedimentary features, including seabed excavation of underlying indurated and semi-consolidated sediments, injection of liquefied muds, microbreccia and spherule-bearing material into fractures in underlying rock, debris flow, turbidity currents, turbulence eddies, climbing ripples, switching currents, slumping, and detachment of blocks (Hassler and Simonson 2001; Glikson et al. 2004) (Figs. 8.2, 8.5, 8.7, 8.10). Evidence for a deep-water depositional environment of some impact-generated tsunami units arises from the presence of microkrystite spherules in rudite units intercalated with low-energy below-wave base pelagic sequences consisting of chert, carbonate, siltstone, and banded ironstones. By contrast impact spherules are less likely to be preserved in shallow water environments where current winnowing and reworking effects tend to destroy the delicate originally glassy spherules. Thus to date no impact fallout units or tsunami deposits were found to be associated with stromatolitic reefs of the Carawine Dolomite, possibly due to reworking and winnowing in shallow-water environments. These relations provide evidence for the effects of short-lived perturbations by large-amplitude tsunami waves followed by little reworking of the below-wave base (below ~200 m) microkrystite spherule-bearing arenite and rudite deposits. Whereas seismic earthquakes of endogenic origin and consequent tsunami waves may, in principle, result in below-wave base tsunami deposits, the unique occurrence of microkrystite spherules in such rudite deposits is diagnostic of impact-generated tsunami events. The tsunami may not have necessarily originated from the impact crater(s), as impact-triggered seismic activity could reactivate faults and plate margins where tsunami waves originated. In the Carawine mega-breccia a high intensity of the tsunami disrupted the underlying partly indurated to indurated carbonate and chert substratum, accompanied by hydraulic fracturing and injection of muds into fractures under high hydraulic pressures, allowing preservation of intact spherules within the liquefied microbreccia and muds.

Significant similarities exist between Pilbara tsunami deposits, ~580 Ma-old impact-triggered tsunami deposits in the Flinders Ranges and the Officer Basin (Wallace et al. 1996), and K-T boundary tsunami deposits in Guatemala and southern Mexico (Keller 2005). These units are marked by turbidites accompanied by exotic fragments and boulders. However, to date few analogues have been reported for autochthonous mega-breccia units similar to the Carawine megabreccia. A likely exception is the spherule-bearing late Devonian Alamo Breccia, which may be associated with a buried impact structure (Warme and Kuehner 1998). Estimates of the energy released by tsunami waves, based on seabed excavation, may be used to infer the amplitude of seismic waves and thus of the parental impacts and of consequent faulting. Such estimates may be coupled with estimates of the diameter of the projectile based on geochemical mass balance calculations (Byerly and Lowe 1994; Shukolyukov et al. 2000; Glikson and Allen 2004) and spherule diameters (Melosh and Vickery 1991).

8.1.5 Impact Ejecta Units and Iron-Rich Sediments

The temporal association between mid-Archaean to early Proterozoic asteroid impact ejecta units and overlying banded iron formations and ferruginous shale suggests that, in many instances, these impacts were closely followed by significant changes in the nature and composition of source terrains of the sediments. Such associations include (Glikson 2006; Glikson and Vickers 2007).

a. Jaspillite overlying a 3470.1 ± 1.9 Ma impact spherule units and tsunami-type breccia of the Antarctic Chert Member of the Mount Ada Basalt, central Pilbara Craton (Glikson et al. 2004) (Figs. 8.2, 8.3, 8.5, 8.8);
b. Banded iron formation (BIF), jaspillite and ferruginous shale overlying the 3258 ± 3 Ma impact spherule unit in the Barberton greenstone belt (BGB) (Lowe et al. 2003);
c. Iron-rich sediments of the Ulundi Formation overlying a 3243 ± 4 Ma impact unit in the BGB;
d. Units of BIF and ferruginous shale of the Nimingarra Iron Formation and Paddy Market Formation of the Gorge Creek Group, central Pilbara Craton, which occur above 3235 ± 3 Ma felsic volcanics correlated with spherule-bearing sections of the BGB;

e. BIF and ferruginous shale of the Marra Mamba Iron Formation overlying the Jeerinah impact (JIL) layer and tsunami-type deposit (2629 ± 5 Ma) of the Fortescue Group, central Pilbara Craton;
f. Ferruginous rich shale and carbonate overlying the ~2.54 Ga Spherule Marker Bed of the Bee Gorge Member of the Wittenoom Formation (Sect. 8.1.3.2; Table 1.2), (Fig. 8.8).

The time interval between the microkrystite spherule units and overlying iron-rich sediments is not always known, thereby rendering genetic relations uncertain. The juxtaposition between impact ejecta units and overlying BIF and jaspillite may be accidental or, alternatively, hint at enrichment of seawater in soluble iron under low-oxygen atmosphere and hydrosphere conditions. In this model enhanced denudation of mafic volcanic terrains uplifted and exposed by impact-induced tectonic movements, or impact-triggered mafic volcanism and hydrothermal activity, led to iron enrichment of sea water. The common presence of BIF above impact ejecta units testifies to below-wave base deposition of the spherules since colloidal/chemical precipitation of iron is retarded in high energy current-dominated environments. A systematic association between impact ejecta and iron-rich sediments may yield useful stratigraphic tracers in the search for impact signatures in early Precambrian terrains.

References

Addison WD, Brumpton GR, Vallini DA, McNaughton NJ, Davis DW, Kissin SA, Fralick PW, Hammond AL (2005) Discovery of distal ejecta from the 1,850 Ma sudbury impact event. Geol 33:193–196

Arndt NT, Nelson DR, Compston W, Trendall AF, Thorne AM (1991) The age of the fortescue group hamersley basin Western Australia from ion microprobe zircon U-Pb results. Aust J Earth Sci 38:261–281

Brauhart C, Groves DI, Morant P (1998) Regional alteration systems associated with volcano-genic massive sulphide mineralization at panorama Pilbara Western Australia. Econ Geol 93:292–302

Buick R, Brauhart CAW, Morant P, Thornett JR, Maniew JG, Archibald JG, Doepel MG, Fletcher IR, Pickard AL, Smith JB, Barley MB, McNaughton NJ, Groves DI (2002) Geochronology and stratigraphic relations of the Sulphur Springs Group and Strelley Granite: a temporally distinct igneous province in the Archaean Pilbara Craton Australia. Precamb Res 114:87–120

BVTP (1981) Basaltic volcanism of the terrestrial planets. Pergamon, New York

Byerly GR (1999) Komatiites of the Mendon Formation: late stage ultramafic volcanism in the Barberton greenstone belt. Spec Pap Geol Soc Am 329:1–29

Byerly GR, Lowe DR (1994) Spinels from Archaean impact spherules. Geochim et Cosmochim Acta 58:3469–3486

Byerly GR, Lowe DR, Wooden JL, Xie X (2002) A meteorite impact layer 3470 Ma from the Pilbara and Kaapvaal Cratons. Science 297:1325–1327

Cannon WF, Schulz KJ, Wright J, Horton D, Kring A (2010) The Sudbury impact layer in the Paleoproterozoic iron ranges of northern Michigan, USA. Geol Soc Am Bull 122:50–75

Chadwick B, Claeys P, Simonson BM (2000) New evidence for a large Palaeoproterozoic impact Spherules in a dolomite layer in the Ketilidian orogen South Greenland. J Geol Soc London 158:331–340

Culler TS, Becker TA, Muller RA, Renne PR (2000) Lunar impact history from 39Ar/40Ar dating of glass spherules. Science 287:1785–1789

Dunlop JSR, Buick R (1981) Archaean epiclastic sediments derived from mafic volcanics North Pole Pilbara Block Western Australia. Geol Soc Aust Sp Publ 7:225–233

Glass BP, Burns CA (1988) In: Microkrystites: a new term for impact-produced glassy spherules containing primary crystallites Proceedings of Lunar Planet Sci Conference XVIII pp 455–458

Glikson AY (1979) Early precambrian tonalite—trondhjemite sialic nuclei. Earth Sci Rev 15:1–73

Glikson AY (1984) Significance of early Archaean mafic–ultramafic xenolith patterns. In: Kroner A, Goodwin AM, Hanson GN (eds) Archaean geochemistry. Springer, Berlin pp 263–280

Glikson AY (2001) The astronomical connection of terrestrial evolution crustal effects of post-3.8 Ga mega-impact clusters and evidence for major 3.2 Ga bombardment of the Earth-Moon system. J Geodyn 32:205–229

Glikson AY, Allen C (2004) Iridium anomalies and fractionated siderophile element patterns in impact ejecta, Brockman Iron Formation, Hamersley Basin, Western Australia: evidence for a major asteroid impact in simatic crustal regions of the early Proterozoic earth. Earth Planet Sci Lett 20:247–264

Glikson AY (2005) Asteroid/comet impact clusters flood basalts and mass extinctions: significance of isotopic age overlaps. Earth Planet Sci Lett 236:933–937

Glikson AY (2004) Bedout: a possible end-permian impact crater offshore of northwestern Australia. Science 306:613

Glikson AY (2006) Asteroid impact ejecta units overlain by iron rich sediments in 3.5–2.4 Ga terrains Pilbara and Kaapvaal cratons: accidental or cause–effect relationships? Earth Planet Sci Lett 246:149–160

Glikson AY (2007) Early Archaean asteroid impacts on Earth: stratigraphic and isotopic age correlations and possible geodynamic consequences. In: Van Kranendonk MJ, Smithies H, Bennett VC (eds) Earth's oldest rocks. Developments in precambrian geology 15

Glikson AY (2007) Siderophile element patterns PGE nuggets and vapor condensation effects in Ni-rich quench chromite-bearing microkrystite spherules 3.24 Ga S3 impact unit. Barberton greenstone belt, Kaapvaal, Craton South Africa Earth Planet Sci Lett 253:1–16

Glikson AY (2008) Field evidence of *Eros*-scale asteroids and impact-forcing of Precambrian geodynamic episodes Kaapvaal (South Africa) and Pilbara (Western Australia) cratons. Earth Planet Sci Lett 267:558–570

Glikson AY, Allen C, Vickers J (2004) Multiple 3.47 Ga-old asteroid impact fallout units Pilbara Craton, Western Australia. Earth Planet Sci Lett 221:383–396

Glikson AY, Hickman AH (1981) Geochemical stratigraphy of Archaean mafic–ultramafic volcanic successions eastern Pilbara Block Western Australia. In: Glover JE, Groves DI (eds) Archaean Geology. Geol Soc Aust Sp Publ 7:287–300

Glikson AY, Vickers J (2005) The 3.26–3.24 Ga Barberton asteroid impact cluster: tests of tectonic and magmatic consequences Pilbara Craton Western Australia. Earth Planet Sci Lett 241:11–20

Glikson AY, Vickers J (2007) Asteroid mega-impacts and Precambrian banded iron formations: 2.63 Ga and 2.56 Ga impact ejecta/fallout at the base of BIF/argillite units Hamersley Basin Pilbara Craton Western Australia. Earth Planet Sci Lett 254:214–226

Hassler SW, Robey HF, Simonson BM (2000) Bedforms produced by impact-generated tsunami, ~2.6 Ga Hamersley Basin. West Austral Geo 135:283–294

Hassler SW, Simonson BM (2001) The sedimentary record of extraterrestrial impacts in deep shelf environments Evidence from the early Precambrian. J Geol 109:1–19

Hassler SW, Simonson BM, Sumner DY, Bodin L (2011) Paraburdoo spherule layer, Hamersley Basin, Western Australia: Distal ejecta from a fourth large impact near the Archaean-Proterozoic boundary. Geology 39:307–310

Hickman AH (1981) Crustal evolution of the Pilbara Block Western Australia. Geol Soc Aust Sp Publ 7:57–69

Jahn B, Simonson BM (1995) Carbonate Pb–Pb ages of the Wittenoom Formation and Carawine Dolomite Hamersley Basin Western Australia, with implications for their correlation with the Transvaal Dolomite of South Africa. Precambr Res 72:247–261

Jirsa MA, Weiblen PW, Vislova T, McSwiggen PL (2008) Sudbury impactite layer near Gunflint Lake, NE Minnesota. Instit Lake Superior Geol Proc 54:42–43

Keller G (2005) Impacts volcanism and mass extinction: random coincidence or cause and effect? Aust J Earth Sci 52:725–757

Kyte FT, Zhou L, Lowe DR (1992) Noble metal abundances in an early Archaean impact deposit. Geochim Cosmochim Acta 56:1365–1372

Kyte FT, Shukolyukov A, Lugmair GW, Lowe DR, Byerly GR (2003) Early Archaean spherule beds: chromium isotopes confirm origin through multiple impacts of projectiles of carbonaceous chondrite type. Geology 31:283–286

LaBerge GL (1966) Altered pyroclastic rocks in iron formation in the hamersley range Western Australia. Econ Geol 61:147–161

Lowe DR (1980) Archaean sedimentation. Ann Rev Earth Planet Sci 8:145–167

Lowe DR, Byerly GR (1986) Early Archean silicate spherules of probable impact origin South Africa and Western Australia. Geology 14:83–86

Lowe DR, Byerly GR (1990) Direct determination of the environmental effects of large meteorite impacts on the Archaean Earth. EOS (Trans Am Geophys Union) 71:1429–1430

Lowe DR, Byerly GR (2010) Did the LHB end not with a bang but with a whimper? 41st Lunar Planet Sci Conf 2563pdf

Lowe DR, Byerly GR (1999) Stratigraphy of the west-central part of the Barberton Greenstone Belt, South Africa. In: Lowe DR, Byerly GR (eds) Geologic evolution of the Barberton Greenstone Belt, South Africa, Geol Soc Am Sp Pap 329:1–36

Lowe DR, Byerly GR, Asaro F, Kyte FJ (1989) Geological and geochemical record of 3400 million year old terrestrial meteorite impacts. Science 245:959–962

Lowe DR, Byerly GR, Kyte FT, Shukolyukov A, Asaro F, Krull A (2003) Spherule beds 3.47–3.24 billion years old in the barberton greenstone belt, South Africa: a record of large meteorite impacts and their influence on early crustal and biological evolution. Astrobiology 3:7–48

Melosh HJ, Vickery AM (1991) Melt droplet formation in energetic impact events. Nature 350:494–497

Mojzsis SJ, Harrison TM (2002) Establishment of a 3.83 Ga magmatic age for the akilia tonalite Southern West Greenland. Earth Planet Sci Lett 202:563–576

Nelson DR, Trendall AF, Altermann W (1999) Chronological relations between the Pilbara and Kaapvaal Cratons. Precam Res 97:165–189

Nutman AP, Friend CRL, Horie K, Hidaka H (2007) The Itsaq gneiss complex of southwestern Greenland and the construction of Eoarchaen crust at convergent plate boundaries. In: Van Kranendonk MJ, Smithies RH, Bennett VC (eds) Earth's oldest rocks, Developments in Precambrian geologyvol 15. Elsevier, Amsterdam pp 187–218

Pope KO, Baines KH, Ocampo AC, Ivanov BA (1997) Energy volatile production and climatic effects of the Chicxulub Cretaceous/Tertiary impact. J Geophys Res 102:21645–21664

Ringwood AE (1986) Origin of the Earth and Moon. Nature 322:323–328

Ryder G (1990) Lunar samples lunar accretion and the early bombardment of the Moon. Eos (Trans Am Geophys Union) 71:313–322

Ryder G (1991) Accretion and bombardment in the Earth–Moon system: the lunar record. Lunar Planet Sci Instit Contrib 746:42–43

Ryder G (1997) Coincidence in the time of the imbrium basin impact and Apollo 15 Kreep volcanic series: impact induced melting? Lunar Planet Sci Instit Contrib 790:61–62

Shukolyukov A, Kyte FT, Lugmair GW, Lowe DR, Byerly GR (2000) The oldest impact deposits on Earth. In: Koeberl C, Gilmour I (eds) Lecture notes in Earth science 92: impacts and the early Earth, Springer, Berlin pp 99–116

Simonson BM (1992) Geological evidence for an early Precambrian microtektite strewn field in the Hamersley Basin of Western Australia. Geol Soc Am Bull 104:829–839

Simonson BM, Cardiff M, Schubel KA (2001) New evidence that a spherule layer in the late Archaean jeerinah formation of Western Australia was produced by a major impact. 32nd Lunar Planet Sci Conf Abstracts, Lunar Planet Instit Contrib 1080, Houston

Simonson BM, Davies D, Hassler SW (2000) Discovery of a layer of probable impact melt spherules in the late Archaean Jeerinah Formation, Fortescue Group, Western Australia. Aust J Earth Sci 47:315–325

Simonson BM, Davies D, Wallace M, Reeves S, Hassler SW (1998) Iridium anomaly but no shocked quartz from late Archaean microkrystite layer: oceanic impact ejecta? Geology 26:195–198

Simonson BM, Hassler SW (1997) Revised correlations in the early precambrian hamersley basin based on a horizon of resedimented impact spherules. Aust J Earth Sci 44:37–48

Simonson BM, Hassler SW, Beukes N (1999) Late Archaean impact spherule layer in South Africa that may correlate with a layer in Western Australia. In: Dressler BO, Sharpton VL (eds) Impact cratering planetary evolution. Geol Soc Am Sp Pap 339, Boulder CO, pp 249–262

Simonson BM, Glass BP (2004) Spherule layers—records of ancient impacts. Ann Rev Earth Planet Sci 32:329–361

Simonson BM, Sumner DY, Beukes NJ, Johnson S, Gutzmer J (2009) Correlating multiple Neoarchean–Paleoproterozoic impact spherule layers between South Africa and Western Australia. Precamb Res 169:100–111

Simonson BM, Hassler SW, Beukes NJ, Sumner DY (2010) Large impacts around the Archaean-Proterozoic boundary—an update. 41st Lunar Planet Sci Conf, 2386.pdf

Trendall AF, Blockley JG (1970) The iron formations of the precambrian hamersley group Western Australia. Geol Surv West Aust Bull 119:365

Trendall AF, Compspton W, Nelson DR, deLaeter JR, Bennett VC (2004) SHRIMP zircon ages constraining the depositional chronology of the hamersley group Western Australia. Aust J Earth Sci 51:621–644

Trendall AF, Nelson DR, deLaeter JR, Hassler SW (1998) Precise zircon U-Pb ages from the marra mamba Iron formation and wittenoom formation, hamersley group, Western Australia. Aust J Earth Sci 45:137–142

Van Kranendonk MJ (2000) Geology of the North Shaw 1:100 000 Sheet. Geol Surv West Australia 1:100 000 Geol Series, p 86

Van Kranendonk MJ, Hickman AH, Smithies RS, Nelson DR (2002) Geology and tectonic evolution of the Archaean North Pilbara Terrain Pilbara Craton, Western Australia. Econ Geol 97:695–732

Vearncombe S, Vearncombe JR, Barley ME (1998) Fault and stratigraphic controls of volcanogenic massive sulphide deposits in the strelley belt, Pilbara Craton, Western Australia. Precamb Res 88:67–82

Wallace MW, Gostin VA, Keays RR (1996) Sedimentology of the Neoproterozoic Acraman impact-ejecta horizon South Australia. Aust Geol Surv Org J Aust Geol Geophys 16:443–451

Warme JE, Kuehner HC (1998) Anatomy of an anomaly: the devonian catastrophic alamo impact breccia of Southern Nevada. Inter Geol Rev 40:189–216

Williams IR (2003) Geology of the Yilgalong 1100 000 Sheet, Western Australian. Geol Surv West Aust 1:100 000 map series

Woodhead JD, Hergt JM, Simonson BM (1998) Isotopic dating of an Archaean bolide impact horizon, Hamersley Basin, Western Australia. Geology 26:47–50

Chapter 9
Large (>100 km Diameter) Impact Structures

Abstract This chapter present summaries of the structure and shock metamorphic features of some of the largest recorded impact structures, including Maniitsoq (southwest Greenland), Yarrabubba (Western Australia), Vredefort (South Africa), Sudbury (Ontario) and Chicxulub (Yucatan, Mexico). Large impact structures (D ≥ 100 km) are more likely to have major seismic, tectonic and magmatic effects.

9.1 Maniitsoq, Southwest Greenland (~2.975 Ga)

The Maniitsoq impact structure in southwest Greenland (Garde et al. 2012) constitutes a >100 km-diameter deformed zone corresponding to a large aero-magnetic anomaly within high grade Archaean metamorphic terrain, including a central 35 × 50 km-large core of crushed re-heated gneisses and comminuted quartzo-feldspathic material. The deformed aureole is cut by widespread fractures, intense fracture cleavage and breccia. Diagnostic planar deformation features in quartz confirm a shock metamorphic origin of the structure (Garde and Glikson 2011). The PDF display multiple intersecting planar sets including Miller indices {0001}, {10–14}, {10–13}, {10–12}, {10–11}, {10–22}, {11–21} and {31–41}, with narrow planar spacing of ~2–5 microns. Quartz grains may be re-deformed and display undulose extinction and curved or kinked PDFs which preserve near-original crystallographic orientations. Other features include fluidization of micro-breccia, evidence of direct K-feldspar and plagioclase melting in migmatized rocks and formation of planar elements in quartz and plagioclase. Regional-scale hydrothermal alteration took place under amphibolite-facies conditions. The structure is cut by ultramafic intrusions showing evidence of crustal contamination, yielding a U–Pb zircon age of 2975 ± 6 Ma. The rocks considered to be exhumed from a depth of 20–25 km below the surface. To date no fallout ejecta corresponding to the Maniitsoq impact has been identified in Archaean terrains.

A. Y. Glikson, *The Asteroid Impact Connection of Planetary Evolution*,
SpringerBriefs in Earth Sciences, DOI: 10.1007/978-94-007-6328-9_9,
© The Author(s) 2013

9.2 Vredefort, Free State, South Africa (2.023 Ga)

The Vredefort structure, the largest impact structure identified on Earth to date, first proposed by Hargraves (1961) and Dietz (1961), comprises an 80-90 km-diameter central plug consisting of Archaean gneisses surrounded by an annular ring of vertically unfolded Proterozoic sediments with a total outer diameter of ~300 km (Fig. 9.1) (Therriault et al. 1997; Grieve et al. 2008) or 280 km (Henkel and Reimold 1998), considered to be eroded to a depth of approximately >5–10 km (McCarthy et al. 1990; Gibson and Reimold 2001; Reimold and Gibson 2006). The vertical uplift of the central core is estimated as 20–30 km (Therriault et al. 1997), accounting for the removal of all deposits of the original transient crater. The basement plug is intersected by nine several km-long radial dykes of impact melt bodies, termed Vredefort Granophyre, as well as breccia-bearing pseudotachylite vein systems which occupy radial fractures and mylonitic zones concentrated along the margin (Buchanan and Reimold 2002). Kamo et al. (1996) determined a 2020 ± 5 Ma U–Pb zircon age for the granophyre. Deformation due to compression in the NW–SE direction and deposition of the Karoo Supergroup over the southeast part of the structure post-date the impact. Evidence of shock metamorphism includes shatter cones (Hargraves 1961), planar deformation features (Leroux et al. 1994), shock deformation in zircon (Kamo et al. 1996), coesite and stishovite (Martini 1978) and extra-terrestrial geochemical signatures

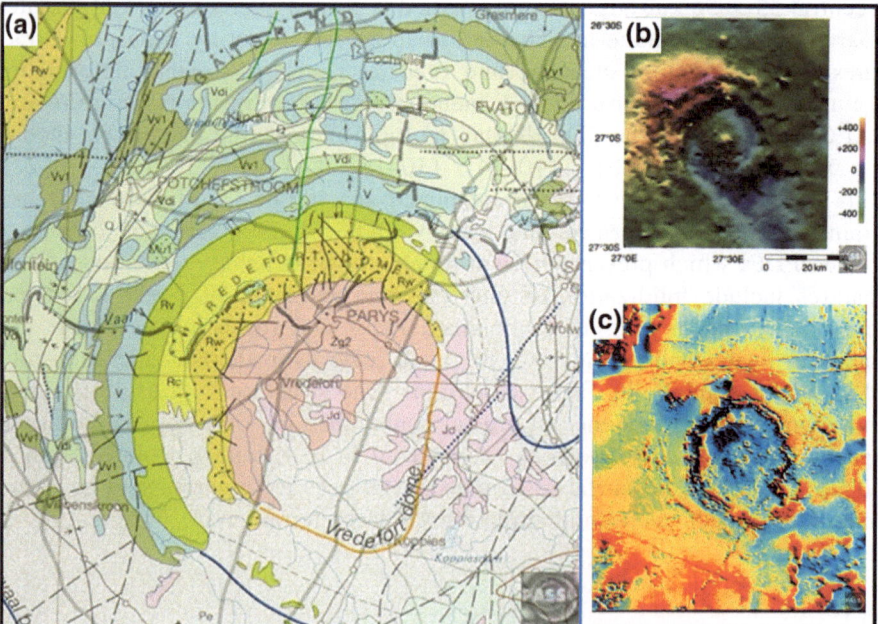

Fig. 9.1 The Vredefort impact structure: **a** Geological sketch map; **b** Bouguer gravity anomaly; **c** Airborne magnetic image (Earth Impact Database, with permission)

(Koeberl et al. 1996). Trace element studies of the pseudotachylite indicate significant enrichment in Iridium over other Vredefort granitic rocks (French et al. 1988). The central dome is enveloped by concentric folds and widespread occurrences of breccia and pseudotachylite veins which extend as far as the northern and northwest sector of the Witwatersrand Basin (McCarthy et al. 1986, 1990; Killick et al. 1988; Reimold and Gibson 2001). The collar of up-folded supracrustal rocks outside the central plug displays outward-directed thrusting whereas outermost sectors display inward-directed normal faulting as far as some 60 km north of the basement core. Seismic data identifies deformed circular sectors around the entire structure (Brink et al. (1997). Lana et al. (2003) suggested that displacements within the central granitoid core occurred by mm to cm-scale differential rotations and slip and by movements along zones of pseudotachylite. Gibson and Reimold (2001) identified paleo-temperatures ranging from 700 to 1,000 °C in the inner core to ~300 °C in the outer collar zone, reflecting pre-impact geothermal gradient, preceding the shock heating.

9.3 Sudbury Impact Structure, Ontario, Canada (~1.85 Ga)

The Sudbury impact structure (Dietz 1964) dated as 1,850 ± 1 Ma (Krogh et al. 1984), comprises a ~60 × 27-km-large, 2.5–3.0 km-thick differentiated layered igneous sheet which includes granodiorite, granophyre, quartz gabbro and norite, termed Sudbury Igneous Complex (SIC) (Fig. 9.2). The SIC is set within fractured and brecciated Archaean basement terrain and is eroded to a depth of about 5–6 km (Dietz 1968; Naldrett et al. 1970; French 1968; Giblin 1984; Pye et al. 1984; Grieve et al. 1991; Therriault et al. 2002; Naldrett 2003; Grieve 2006); Thompson et al.

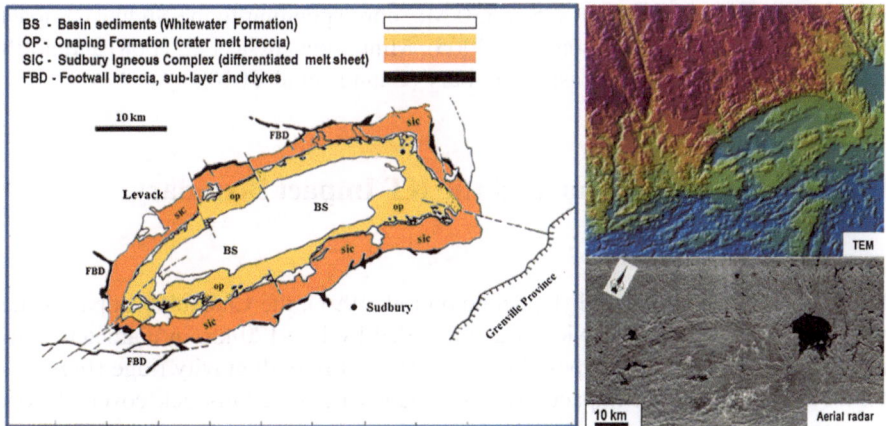

Fig. 9.2 The Sudbury impact structure: **a** Geological sketch map (Elsevier, with permission); **b** Digital Elevation model; **c** Aerial Radar (Earth Impacts Database, with permission)

1998). Post-impact deformation, represented by NW–SE shortening during the Penokean orogeny (Rousell 1984; Riller 2005), resulted in preservation of the entire sequence from impacted basement, to the SIC, to overlying melt breccia (Onaping Formation) and black shale, as well as radial dikes of impact melt of an overall noritic composition (Tuchscherer and Spray 2002). An overall diameter of ~260 km is suggested from surface to geophysical data, including a N–S LITHOPROBE geophysical transect (Tuchscherer and Spray 2002; Spray et al. 2004; Boerner et al. 2000), disclosing thrust faults and ductile shears related to post-impact deformation. Evidence of shock metamorphism is provided by planar deformation features in quartz and feldspar in basement rocks and the overlying Onaping Formation (French 1968; Dence 1972) and by shatter cones (Guy-Bray 1966). Breccia occurrences are concentrated about 5–15 km from the SIC but are also found as far as 80 km from the SIC (Spray 1997). Spray et al. (2004) delineated four breccia rings at 90, 130, 180 and 260 km north of the SIC corresponding to LANDSAT lineament analysis by Butler (1994). Whether these observations allow definition of Sudbury as a multi-ring impact structure remains uncertain (Grieve et al. 2008).

A hydrous composition of parent melt of the SIC is suggested by the dominance of amphibole and biotite, granophyre and deuteric alteration. Segregations of sulfide precipitated from the melt and from hydrothermal fluids represent advanced magmatic differentiation of the SIC. Differences between the SIC and layered intrusions elsewhere include its intermediate composition, hydrous nature, crustal/granitic isotopic values, presence of normative corundum, prevalence of granophyre, presence of plagioclase xenocrysts showing complex twinning and planar deformation features and presence of PDF in quartz xenocrysts (Therriault et al. 2002). The SIC is overlain in part by 1.4–1.6 km thick breccia and melt breccia of the Onaping Formation, containing shock metamorphosed and hydrothermally altered lithic clasts and glass fragments and flows, analogous to suevite breccia of the Ries Crater (Grieve et al. 2010). Allochtonous fragmental ejecta containing shock-induced planar deformation features in quartz and altered devitrified glass and/or accretionary lapilli and dated within the 1,875–1,830 Ma, corresponding to the Sudbury impact, is reported in Ontario (Addison et al. 2005), Minnesota (Jirsa et al. 2008) and northern Michigan, 500–700 km east of Sudbury (Cannon et al. 2010).

9.4   Chicxulub, Yucatan, and the KT Impact Boundary (65 Ma)

Discovered by Glen Penfield (Hildebrand et al. 1991), the Chicxulub impact structure, Yucatan Peninsula, Mexico, was identified by its ~170 km in diameter multiring Bouguer anomaly superposed on an older north–south gravity ridge (Figs. 5.1, 9.3). As a subterranean impact structure underlying a ~2 km-thick cover of post-impact Tertiary sediments, evidence for the Chicxulub impact structure hinges on geophysical data and drill cores (Sharpton et al. 1993; Pilkington et al. 1994; Hildebrand et al. 1995). Three concentric magnetic zones correspond to the gravity anomaly. An inner high-amplitude (>500 nT) anomaly ~40 km-diameter zone coincides with a gravity high and a high seismic velocity, reflecting crystalline basement uplift ~3 km of the surface (Pilkington and Hildebrand 2000). Deep-seated sectors with seismic velocity of 6.0 –6.3 km/s are interpreted in terms of incorporation of crystalline basement. The inner gravity ring surrounds a strong magnetic high. An outer ~90 km-diameter magnetic zone overlaps an inward dipping seismic reflector which forms a boundary between inner brecciated zone and an outer more intact zone (Morgan et al. 2000). On land the structure is expressed by a 160 km-diameter ring of Cenote sinkholes (Pope et al. 1993). Off shore seismic reflection profiles and the Tertiary cover deepen at a diameter of about 180 km. Outer limits of the structure are indicated by a marine seismic reflection survey, defining a peak ring surrounded by circular faults at diameters of ~195 and ~240 km (Morgan et al. 1997). The peak ring rises approximately 600 meters above the apparent crater floor (Morgan et al. 1997) and is underlain by a steeply dipping low velocity seismic zone interpreted as injected breccia overlying fractured crystalline basement (Vermeesch and Morgan 2004). Combined seismic data and drill holes indicate a thickening of impactite deposits around the peak ring (Hildebrand et al. 1998; Morgan et al. 2000). Drilling outside the peak ring intersected approximately 490 m of melt breccia overlying brecciated dolomite. The Impact breccia displays shock metamorphic features and impact melting (Sharpton et al. 1996; Stöffler et al. 2004). Ar–Ar ages on melt fractions and on tektite suggest a 64.98 ± 0.05 Ma (Swisher et al. 1992).

Outcrops of pelagic sediments of early Jurassic to Oligocene age (~185–30 Ma) in the Umbrian Apennines, Italy, contain a para conformity representing a hiatus across the Cretaceous-Paleocene boundary, across which underlying a foraminifera-rich white limestone facies containing large-scale Globotruncana *contusa* is abruptly replaced by overlying clay-rich red limestone termed Scaglia rossa with smaller foraminifera (Globigerina *eugubina*) and micron-scale algal coccoliths (Alvarez et al. 1980) (Fig. 9.4). At the classic Gubbio locality a ~1 cm-thick boundary clay layer consists of a lower ~5 mm-thick grey clay zone dominated by clastic material and an upper ~5 mm-thick red clay zone, where the red clay zone termed 'fire layer' contains an Iridium anomaly of up to ~9 ppb. The boundary coincides with a major geomagnetic reversal. The geomagnetic reversal is correlated with the marine magnetic anomaly sequence dated with foraminifera. The classic KT boundary ejecta outcrop at Gubbio, Italy, is portrayed in Fig. 9.4. KT boundary impact ejecta, best preserved in deep water environments, have been identified along the Maastrichtian—Danian boundary in more than 101 sites and Iridium anomaly in over 85 sites around the globe (Claeys et al. 2002). The ejecta coincide with erosion of Maastrichtian sediments and onset of clastic sediments and breccia attributable to seismic and tsunami effects around the Gulf of Mexico and the Atlantic Ocean. The grain size of quartz with planar deformation features increases westward of Chicxulub (Alvarez et al. 1995).

A wide spectrum of studies following the KT impact boundary discovery established the environmental effects of mega-impact, including atmospheric flash igniting forest fires, bolide penetration and explosion, seismic mega-earthquakes with possible faulting, cratering, crustal rebound, fragmentation, fusion,

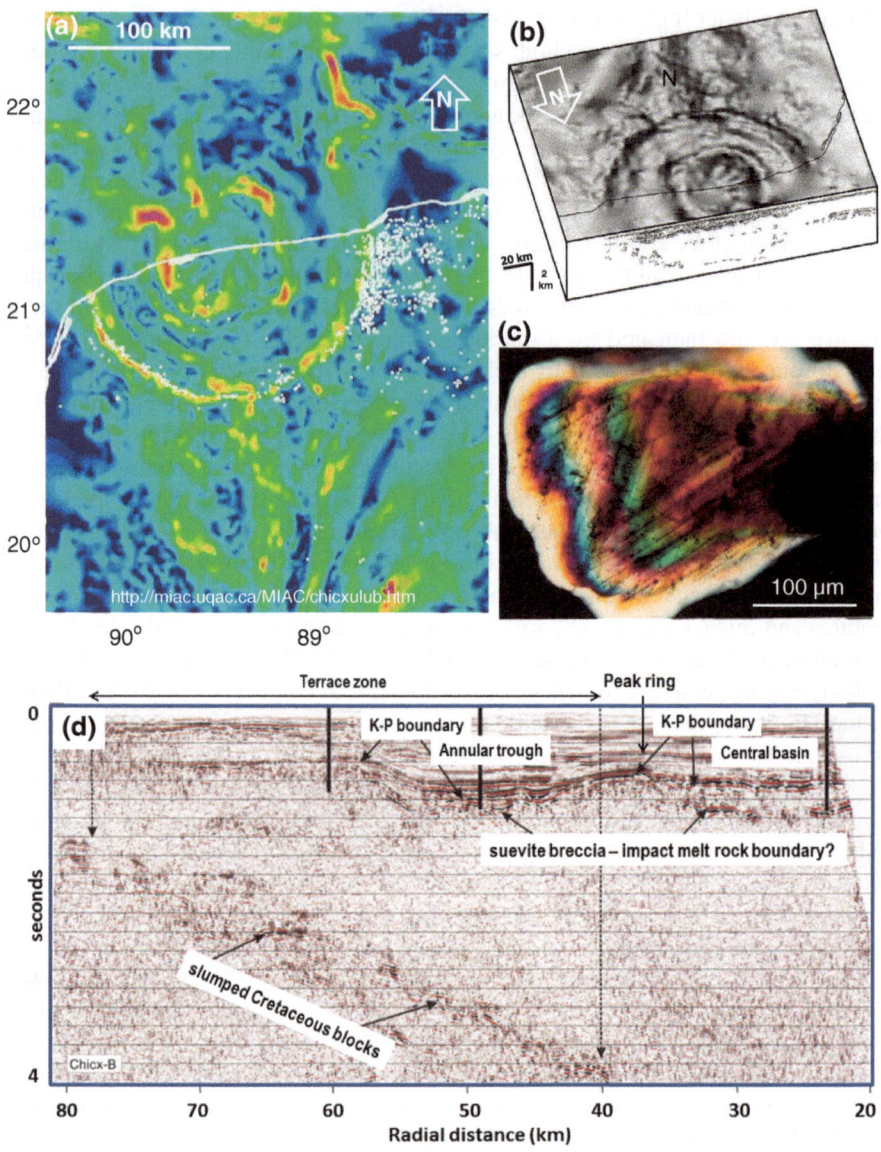

vaporization of rocks, an ejecta blanket, atmospheric and stratospheric dusting, vapor clouding, atmospheric chemical reactions, acid rain, destruction of ozone, release of CO_2 from target materials and ensuing greenhouse effects, biological mass extinction, as well as possible volcanic consequences. Aerosol clouding reducing sunlight and thereby arresting photosynthesis would attack the root of the terrestrial and marine good chain, followed by extreme warming induced by the released greenhouse gases. According to Beerling et al. (2002) Late Cretaceous background CO_2 levels of 350–500 ppm increased to at least 2,300 ppm within 10,000 years of the KT impact, due to instantaneous transfer of approximately

◀**Fig. 9.3** The Chicxulub impact structure, northern Yucatan Peninsula, Mexico. From A. Hildebrand http://miac.uqac.ca/MIAC/chicxulub.htm by permission. **a** Horizontal gradient map of a Bouguer gravity anomaly over the Chicxulub crater, constructed from gravity measurements taken in 1948. Most of the concentric gradient features can be related to inferred structural elements of the buried crater, including the central uplift (note the radial features revealed in the uplift), the collapsed transient cavity edge, faults in the zone of slumping, and the edge of the topographic basin - the now buried crater. White dots represent the locations of water-filled sink-holes, solution collapse features called cenote common in the limestone rocks of the region; **b** A perspective plot of the ~180 km-diameter Chicxulub crater with a cutaway view showing a cross section of the crater as revealed by seismic reflection data. This view is looking to the south; the Yucatán coastline is shown by a thin dark line. Note the different horizontal and vertical scales such that the cross section has a vertical exaggeration of approximately ten times. The vertical scale is approximate as seismic velocities vary with depth and rock type. The interpretation of seismic reflection data shown on the face of the cutaway is from Camargo and Suarez (1994). The seismic data reveal much of the crater's structure including the ~1 km-thick Tertiary sediments filling the crater, the crater edges, the peak ring, and the down-dropped blocks in the crater's zone of slumping; **c** A 0.32 mm shocked quartz grain from intra-crater breccia sample from the Yucatán-6 hole located ~50 km from the crater's center and which penetrated ~500 meters of impact melt and breccia. The quartz grain displays dislocations along preferred crystallographic orientations along at least 8 sets of planar deformation features. The lamellae are decorated with inclusions. Impact is the only natural process known to produce shock waves of sufficient strength to cause deformation of this type; **d** Marine reflection seismic profile Chicx-B East-West cross section correlated with structural and lithological elements. This material is reproduced with permission of John Wiley & Sons, Inc.

Fig. 9.4 The KT impact boundary at Gubbio, Apennines, Italy: **a** showing a plankton-rich limestone bed, overlain by green–brown fragmental ejecta layer, overlain by a red "fire layer" which contains microkrystite spherules, overlain by 'Scaglia Rossa'—red limestone containing only small foraminifera, **b** a thin section showing the contrast between the white limestone containing large Globotruncana *contusa* and the red Scaglia Rossa containing Globigerina eugubina (courtesy of Alessandro Montanari)

~4,600 billion ton carbon (GtC) with consequent rise in greenhouse radiative forcing of +12 W/m2, raising surface temperatures by ~7.5 °C, accounting for the mass extinction of ~44 % of genera (Keller 2005).

References

Addison WD, Brumpton GR, Vallini DA, McNaughton NJ, Davis DW, Kissin SA, Fralick PW, Hammond AL (2005) Discovery of distal ejecta from the 1850 Ma Sudbury impact event. Geology 33:193–196

Alvarez L, Alvarez W, Asaro F, Michel HV (1980) Extraterrestrial cause for the Cretaceous-Tertiary extinction. Science 208:1095–1108

Alvarez W, Claeys P, Kieffer SW (1995) Emplacement of KT boundary shocked quartz from Chicxulub crater. Science 269:930–935

Beerling DJ, Lomax BH, Royer DL, Upchurch GR, Kump LR (2002) An atmospheric PCO_2 reconstruction across the Cretaceous-Tertiary boundary from leaf mega fossils. Proc Nat Acad Sci 99:7836–7840

Boerner DE, Milkereit B, Davidson A (2000) Geoscience impact: a synthesis of studies of the Sudbury structure. Canadian J Earth Sci 37:477–501

Brink MC, Wanders FB, Bischoff AA (1997) Vredefort: a model for the anatomy of an astroblemes. Tectonophysics 270:83–114

Buchanan PC, Reimold WU (2002) Planar deformation features and impact glass in inclusions from the Vredefort granophyre South Africa. Meteor Planet Sci 37:807–822

Butler HR (1994) Lineament analysis of the Sudbury multiring impact structure. In: Large meteorite impacts and planetary evolution. Geol Soc Am Sp Pap 293:319–330

Camargo ZA, Suarez GR (1994) Evidencia sismica del crater de impacto de Chicxulub. Boletin de la Asociacion Mexicana de Geofisicos de Exploracion 34:1–28

Cannon WF, Schulz KJ, Wright J, Horton D, Kring A (2010) The Sudbury impact layer in the Paleoproterozoic iron ranges of northern Michigan, USA. Geol Soc Am Bull 122:50–75

Claeys P, Kiessling W, Alvarez, W (2002) Distribution of Chicxulub ejecta at the Cretaceous-Tertiary boundary. In: Koeberl C., MacLeod KG (eds) Catastrophic events and mass extinctions: impacts and beyond: Boulder, Colorado, Geol Soc Am Spec Pap 356:55–68

Dence MR (1972) Meteorite impact craters and the structure of the Sudbury basin. Geol Assoc Canada Sp Pap 10:7–18

Dietz RS (1961) Vredefort ring structure: meteorite impact scar? J Geol 69:496–505

Dietz RS (1964) Sudbury structure as an astroblemes. J. Geol 72:412–434

Dietz RS (1968) Shatter cones in cryptoexplosion structures. In: French BM, Short NM (eds) Shock metamorphism of natural materials. Mono Book Corp Baltimore, pp 267–285

French BM (1968) Sudbury structure Ontario: some petrographic evidence for an origin by meteorite impact. In: French BM, Short NM (eds) Shock metamorphism of natural materials. Mono Books, Baltimore, pp 383–412

French BM, Orth CJ, Quintana LR (1988) Iridium in the Vredefort Bronzite Granophyre—impact melting and limits on a possible extraterrestrial component. Lunar and Planetary Science Conference 19th Houston TX 14–18, pp 733–744

Garde AA, Glikson AY (2011) Recognition of re-deformed planar deformation features (PDF) in large impact structures. 74th Ann Meteor Soc Meet, 5246 pdf

Garde AA, McDonald I, Dyck B, Keulen N (2012) Searching for giant ancient impact structures on Earth: the Meso-Archaean Maniitsoq structure, West Greenland. Earth Planet Sci Lett 2012:337–338

Giblin PE (1984) History of exploration and development of geological studies and development of geological concepts. In: Pye E, Naldrett AJ, Giblin PE (eds) The geology and ore deposits of the Sudbury structure. Ontario Geol Surv, pp 3–24

Gibson RL, Reimold WU (2001) The Vredefort impact structure South Africa: The scientific evidence and a two-day excursion guide. Council Geosci Mem 92:111 p

Grieve RAF (2006) Impact structures in Canada. Geol Assoc Canada, p 210

Grieve RAF, Stöffler D, Deutsch A (1991) The sudbury structure: controversial or misunderstood? J Geophys Res 96:22753–22764

Grieve RAF, Reimold WU, Morgan J, Riller U, Pilkington (2008) Observations and interpretations at Vredefort Sudbury and Chicxulub: towards an empirical model of terrestrial impact basin formation. Meteor Planet Sci 43:855–882

Grieve RAF, Ames DE, Morgan JV, Artmieva N (2010) The evolution of the Onaping formation at the Sudbury impact structure. Meteor Planet Sci 45:159–782

Guy-Bray J (1966) Shatter cones at Sudbury. J. Geol 74:243–245

Hargraves RB (1961) Shatter cones in the rocks of the Vredefort ring. Geol Soc S Afr Trans 64:147–154

Henkel H, Reimold WU (1998) Integrated geophysical modeling of a giant complex impact structure: anatomy of the Vredefort structure South Africa. Tectonophysics 287:1–20

Hildebrand AR, Penfield GT, Kring DA, Pilkington M, Camargo ZA, Jacobsen SB, Boynton WV (1991) A possible Cretaceous-Tertiary boundary impact crater on the Yucatan Peninsula, Mexico. Geology 19:867–871

Hildebrand AR, Pilkington M, Connors M, Ortiz-Aleman C, Chavez RE (1995) Size and structure of the Chicxulub crater revealed by horizontal gravity gradients and cenotes. Nature 376:415–417

Hildebrand AR, Pilkington M, Ortiz-Aleman C, Chavez RE, Urrutia-Fucugauchi J, Connors M, Graniel-Castro E, Camarago ZA, Halpenny JF, Niehaus D (1998) Mapping Chicxulub crater structure with gravity and seismic reflection data. In: Meteorites Flux with time and impact effects. Sp Publ Geol Soc London 140:153–173

Jirsa MA, Weiblen PW, Vislova T, McSwiggen PL (2008) Sudbury impactite layer near Gunflint lake, NE Minnesota. Instit Lake Superior Geol Proc 54:42–43

Kamo SL, Reimold WU, Krogh TE, Colliston WP (1996) A 2.023 Ga age for the Vredefort impact event and a first report about shock metamorphosed zircons in pseudotachylitic breccias and granophyre. Earth Planet Sci Lett 144:369–388

Keller G (2005) Impacts volcanism and mass extinction: random coincidence or cause and effect? Aust J Earth Sci 52:725–757

Killick AM, Thaites AM, Germs GJB, Schoch AE (1988) Pseudotachylite associated with a bedding-parallel fault zone between the Witwatersrand and Ventersdorp supergroup South Africa. Geolo Rund 77:329–344

Koeberl C, Reimold WU, Shirley SB (1996) Re-Os isotope and geochemical study of the Vredefort granophyre: clues to the origin of the Vredefort structure South Africa. Geology 24:913–916

Krogh TE, Davis DW, Corfu F (1984) Precise U-Pb zircon and Baddeleyite ages for the Sudbury area.In: The geology and ore deposits of the Sudbury structure. Ontario Geol Surv Sp 1:431–448

Lana C, Gibson RL, Reimold WU (2003) Impact tectonics in the core of the Vredefort Dome South Africa: implications for central uplift formation in very large impact structures. Meteor Planet Sci 38:1093–1107

Leroux H, Reimold WU, Doukhan JC (1994) A TEM investigation of shock metamorphism in quartz from the Vredefort Dome South Africa. Tectonophysics 230:223–239

Martini JEJ (1978) Coesite and stishovite in the Vredefort Dome South Africa. Nature 272:715–717

McCarthy TS, Charlesworth EG, Stanistreet IG (1986) Post-Transvaal structural features of the northern portion of the Witwatersrand basin. Trans Geol Soc S Afr 89:311–324

McCarthy TS, Stanistreet IG, Rob LJ (1990) Geological studies related to the origin of the Witwatersrand basin and its mineralization—an introduction and a strategy for research and exploration. S Afr J Geol 93:1–4

Morgan JV, Warner M, Chicxulub Working Group (1997) Size and morphology of the Chicxulub impact crater. Nature 390:472–476

Morgan JV, Warner MR, Collins GS, Melosh HJ, Christeson GL (2000) Peak ring formation in large impact craters. Earth Planet Sci Lett 183:347–354

Naldrett AJ (2003) From impact to riches: evolution of geological understanding as seen at Sudbury Canada. GSA Today 13:4–9

Naldrett AJ, Bray JG, Gasparrini EL, Podolsky T, Rucklidge JC (1970) Cryptic variation and petrology of the Sudbury Nickel irruptive economic. Geology 65:122–155

Pilkington M, Hildebrand AR (2000) Three-dimensional magnetic imaging of the Chicxulub crater. J Geophys Res 105:23479–23491

Pilkington M, Hildebrand AR, Ortiz-Aleman C (1994) Gravity and magnetic field modeling and structure of the Chicxulub crater, Mexico. J Geophys Res 99:13147–13162

Pope KO, Ocampo AC, Duller CE (1993) Surficial geology of the Chicxulub impact crater Yucatán Mexico. Earth Moon Planets 63:93–104

Pye EG, Naldrett AJ, Giblin PE (eds) (1984) The geology and ore deposits of the Sudbury structure. Ontario Geol Surv Sp Vol 1 Toronto, 604 p

Reimold WU, Gibson RL (2006) The melt rocks of the Vredefort impact structure—Vredefort granophyre and pseudotachylitic breccias: implications for impact cratering and the evolution of the Witwatersrand Basin. Chemie der Erde Geochem 66:1–35

Riller U (2005) Structural characteristics of the Sudbury impact structure Canada: impact induced and orogenic deformation—A review. Meteor Planet Sci 40:1723–1740

Rousell DH (1984) Structural geology of the Sudbury basin. In: Pye E, Naldrett AJ, Giblin PE (eds) The geology and ore deposits of the Sudbury structure. Ontario Geol Surv Sp 1:83–96

Sharpton VL, Burke K, Camargo Z, Hall SA, Lee DS, Marin LE, Suárez R, Quezada M, Spudis PD, Urrutia-Fucugauchi J (1993) Chicxulub multi-ring impact basin: size and other characteristics derived from gravity analysis. Science 261:1564–1567

Sharpton VL, Martin LE, Carney JL, Lees S, Ryder G, Schuraytz BC, Sikora P, Spudis PD (1996) A model of the Chicxulub impact basin based on the evaluation of geophysical data well logs and drill core samples. Geol Soc of Am Sp Pap 307:55–74

Spray JG (1997) Superfaults. Geology 25:627–630

Spray JG, Butler HR, Thompson LM (2004) Tectonic influences on the morphometry of the Sudbury impact structure: implications for terrestrial cratering and modeling. Meteor Planet Sci 39:287–301

Stöffler D, Artemieva A, Ivanov B, Hecht L, Kenkmann T, Schmitt RT, Tagle RA, Wittmann A (2004) Origin and emplacement of the impact formations at Chicxulub Mexico as revealed by the ICDP deep drilling at Yaxcopoil-1 and by numerical modeling. Meteor Planet Sci 39:1035–1067

Swisher CC, Grahales-Nishimura JM, Montanari A, Margolis SV, Claeys Ph, Alvarez W, Renne P, Cedillo-Pardo E, Florentin JM, Maurasse R, Curtis GH, Smit J, McWilliams MO (1992) Coeval 39Ar/40Ar ages of 650 million years ago from Chicxulub crater melt rock and Cretaceous-Tertiary boundary tektites. Science 257:954–958

Therriault AM, Grieve RAF, Reimold WU (1997) Original size of the Vredefort Structure: implications for the geological evolution of the Witwatersrand Basin. Meteor Planet Sci 32:71–77

Therriault AM, Anthony D, Flower R, Grieve RAF (2002) The Sudbury igneous complex: a differentiated impact melt sheet. Econ Geol 97:1521–1540

Thompson LM, Spray JG, Kelley SP (1998) Laser probe argon-40/argon-39 dating of Pseudotachylite from the Sudbury structure: evidence for post-impact thermal overprinting in the North range. Meteor Planet Sci 33:1259–1269

Tuchscherer MG, Spray JG (2002) Geology mineralization and emplacement of the Foy offset dike, Sudbury. Econ Geol 97:1377–1398

Vermeesch PM, Morgan JV (2004) Chicxulub central crater structure: Initial results from physical property measurements and combined velocity and gravity modeling. Meteor Planet Sci 39:1019–1034

Chapter 10
Asteroid Impact Clusters and Isotopic Age Peaks

Abstract This chapter tests possible age correlations between large impact events and impact clusters and tectonic and magmatic events. Whereas, the ages of a number of impact events and geodynamic events correlate within error, the only instance where confident genetic relationships between such events can be established are the 3.26 – 3.24 Ga impact cluster identified in the Barberton Greenstone belt, South Africa, and the abrupt transformation from mafic–ultramafic crust to semi-continental crust in the Kaapvaal and Pilbara cratons.

The origin of isotopic age frequency peaks is generally interpreted in terms of episodic thermal and magmatic events, intermittent plate-tectonics events, mantle dynamics and plume activity (Condie 1995; Davies 1995; Smithies et al. 2005). A compilation of 5,509 U–Pb zircon ages by O'Reilly et al. (2008) defines at least nine prominent age distribution peaks at ~3,336, ~3,212, ~2,675, ~2,560, ~2,030, ~1,625, ~1,165, ~570 and ~290 Ma (Fig. 10.1). This included 1,680 zircon U–Pb age determinations with high initial d^{176}Hf values signifying derivation of magmas from juvenile mantle with frequency peaks at ~3,336, ~3,212, ~2,750, ~2,560, ~1,650, ~1,198 and ~290. Mantle-derived magmas are distinguished from reworked crustal magmas by their high d^{176}Hf values of 0.2807 which reflect a high ^{176}Lu/^{177}Hf of >0.02 (Pietranik et al. 2008). The statistical significance and the origin of these peaks remain to be determined.

As well as furnishing regional and inter-continental stratigraphic markers, impact fallout and associated tsunami units allow intercontinental time-event correlations (Figs. 8.4, 10.2). Examples include:

1. Correlation between ~3,467 ± 4 Ma chert/arenite impact fallout unit in the Pilbara Craton and a ~3,470.4 ± 2.3 Ma unit in the Onverwacht Group, Barberton Mountain Land, Transvaal (Sect. 8.1.1) (Byerly et al. 2002) (Table 1.2).
2. Correlation between the 3,258 – 3,225 Ma impact ejecta units at the base of the Figure Tree Group, Barberton Greenstone Belt (Lowe et al. 2003), and the transition between the 3.24 Ga volcanic Sulphur Springs Group and the Gorge Creek Group in the central Pilbara (Table 1.2).

A. Y. Glikson, *The Asteroid Impact Connection of Planetary Evolution*,
SpringerBriefs in Earth Sciences, DOI: 10.1007/978-94-007-6328-9_10,
© The Author(s) 2013

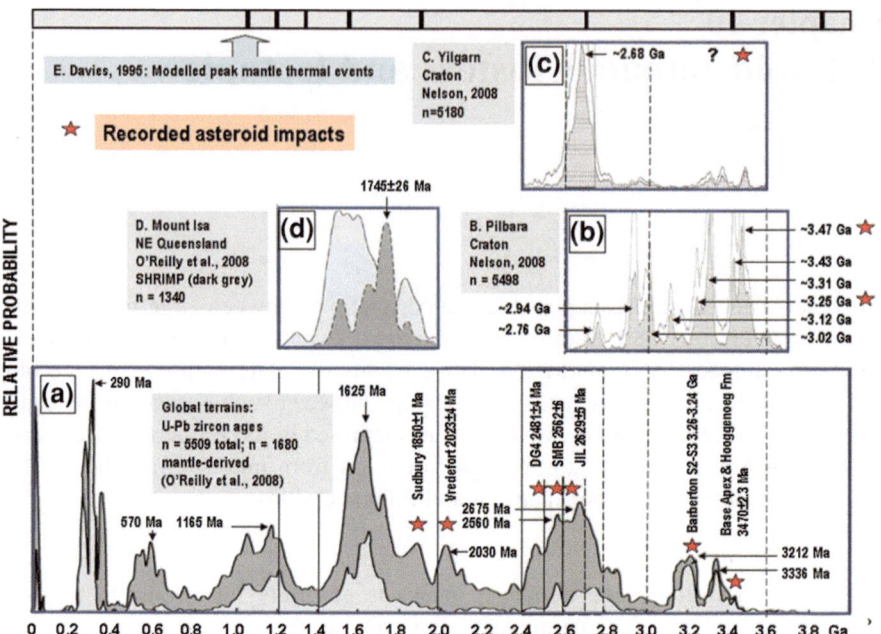

Fig. 10.1 Isotopic U–Pb zircon age frequency distribution (relative probability) diagrams. **a** Global (after O'Reilly et al. 2008). **b** Pilbara Craton, Western Australia (after Nelson 2008). **c** Yilgarn Craton, Western Australia (after Nelson 2008). **d** Northwest Queensland (after Nelson 2008). **e** Model mantle thermal events (Davies 1995). Stars represent recorded large asteroid impacts. Impact ages are located *above stars*; histogram peaks are indicated by *horizontal arrows* (Glikson and Vickers 2010; AJES, Geological Society of Australia, by permission)

3. Possible correlation between the 2,629 ± 5 Ma top Jeerinah Formation impact fallout unit (JIL) and an impact fallout unit in the <2,647 ± 30 Ma Monteville Formation, western Transvaal (Simonson et al. 1999; Simonson and Glass 2004) (Table 1.2, Fig. 10.2).
4. Correlation between the 2,481 ± 4 Ma Dales Gorge spherule bed and a 10 cm-thick spherules and overlying breccia unit at the lower part of the Kuruman Iron Formation, near Griqualand, western Transvaal Basin (Table 1.2, Fig. 10.2).

According to model global thermodynamic peak events by Davies (1995) episodic mantle overturn events were related to temporal cooling, changes in phase-transformation barriers, changes in the mantle layering structure, and subduction of plates. In this model 2.0 – 1.0 Ga phase changes are correlated with peak thermodynamic events, with preferred model peaks at ~3.85, ~3.4, ~2.7, ~1.85, ~1.18 and ~1.05 Ga (Davies 1995, Fig. 5 model [a]). The model peaks correspond in part to the following events:

1. The ~3.85 Ga peak correlates with the end of the Late Heavy Bombardment (Ryder 1990, 1991, 1997);

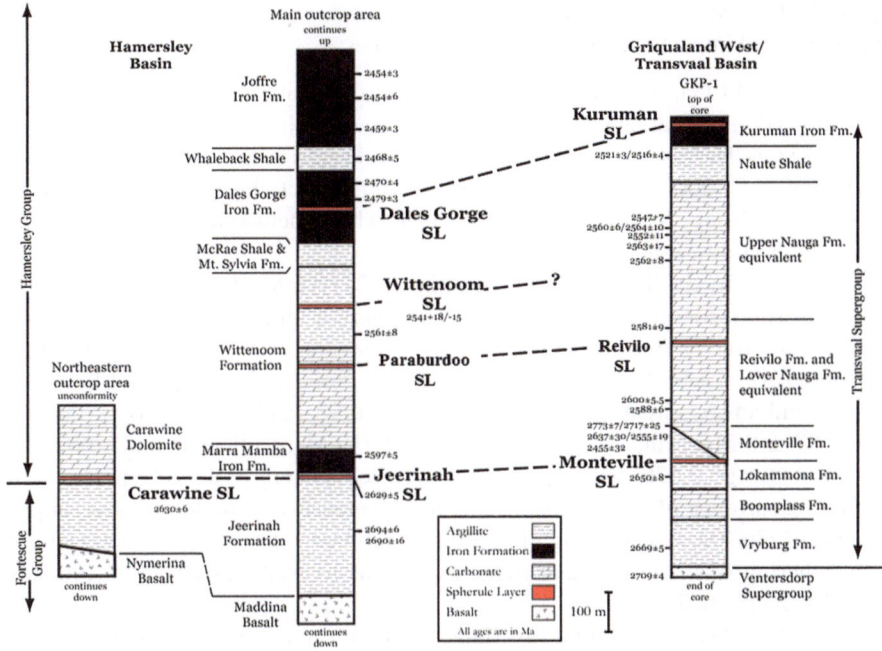

Fig. 10.2 Stratigraphic correlations between the Hamersley Basin, Western Australia, and the Transvaal Basin, South Africa, indicating impact units (in *red*). (Simonson et al. 2010) (Scott Hassler, by permission)

2. The ~3.4 and ~2.7 Ga peaks correlate with periods of maximum volcanic activity in Archaean greenstone belts (Nelson 2008);
3. The ~1.85 Ga peak correlates with the Sudbury mega-impact event (D ~ 250 km) and with thermodynamic events in Proterozoic mobile belts, for example the Capricorn Orogeny, Western Australia (Nelson 2008);
4. The ~1.6 and 1.18 Ga peaks correlate with extensive thermodynamic events in Proterozoic mobile belts, for example in northwestern Queensland and central Australia (O'Reilly et al. 2008).

However U–Pb zircon age peaks at ~3.34, ~3.21, ~2.03, ~0.57 and ~0.29 Ga (Nelson 2008; O'Reilly et al. 2008) are not correlated with Davies' (1995) model (a) but correspond to the following events: (1) the ~3.34 Ga peak correlates with extensive volcanic activity (Euro Basalt, Pilbara Craton: Hickman 2004; Hickman and Van Kranendonk 2004; Van Kranendonk et al. 2007); (2) the ~3.21 Ga peak is within error from the 3.26 – 3.24 asteroid cluster; and (3) the ~2.03 Ga peak overlaps the Vredefort mega-impact (2,023 ± 4 Ma), Kaapvaal Craton, and the Glenburgh Orogeny in Western Australia (Nelson 2008).

Although the observed isotopic age peaks may have arisen from multiple factors, several frequency peaks indicated by compilations of igneous and detrital zircon U–Pb ages (Poujol et al. 2003; Nelson 2008; O'Reilly et al. 2008) overlap major asteroid impact events (Fig. 10.1) as follows:

1. ~3.47 – 3.44 Ga. This period includes peak magmatic events in the Kaapvaal Craton (Poujol et al. 2003) and the Pilbara Craton (Nelson 2008).
2. ~3.212 Ga. This peak (O'Reilly et al. 2008) and peaks in the range ~3.26–3.22 Ga (Poujol et al. 2003) correspond to the asteroid impact cluster at the base of the Fig Tree Group, Barberton greenstone belt, which include impacts at ~3.24 and ~3.26 Ga (Lowe et al. 2003; Glikson 2008).
3. ~2.675 – 2.56 Ga. This peak includes the 2,629 + 5 Ma Jeerinah Impact Layer (JIL) (Simonson et al. 2000), ~2.57 Ga spherule unit and the 2,565 + 9 Ma Spherule Marker Bed, Hamersley Basin (Simonson 1992; Hassler et al. 2011). The ~2.57 – 2.56 Ga impactites overlap the Sm–Nd 2,586 ± 16 Ma age of the Zimbabwe Great Dyke (Mukasa et al. 1998) and is close to U–Pb ages (2,587 ± 8 Ma, Mukasa et al. 1998; 2,578 ± 0.9 Ma, Collerson et al. 2002) and to Re–Os age 2,576 ± 1 Ma of the dyke (Schoenberg et al. 2003). The common occurrence of ferruginous shale and banded iron-formation above impact ejecta units (Sect. 8.1.5) signifies iron enrichment of sea water, likely due to weathering of mafic volcanic rocks as well as hydrothermal activity (Glikson and Vickers 2007).
4. The ~2.48 Ga impact unit overlaps with Huronian mafic igneous activity (2,473 ± 16/79 Ma: Heaman 1997), marking the onset of large-scale mafic dyke systems at 2,473 – 2,446 Ma. The Matachewan–Hearst dyke systems involve volumes of at least 50,000 km^3 of basaltic magma intruding an area of 250,000 km^2 (Hall and Bates 1990; Heaman 1997). Remnants of a global ~2.50 – 2.42 Ga magmatic province are identified in many Archaean cratons, including the ~2.42 Ga Scourie dyke swarm (Scotland), ~2.45 – 2.44 Ga Karelian layered mafic intrusions, flood basalts, and dyke swarms (Finland and Russia), ~2.42 Ga Widgiemooltha dyke swarm and Jimberlana intrusion (Western Australia), dyke systems in the Vestfold Craton (Antarctica) and the ~2.42 Ga Bangalore dyke swarm (Dharwar Craton, India). The temporal coincidence of the ~2.48 Ga Dales Gorge–Kuruman asteroid mega-impact with the onset of global dyking events 2.48 – 2.42 Ga may potentially be interpreted in terms of impact-induced crustal fracturing.

The ~2.03 Ga peak approximates the age of the Vredefort impact structure dated at 2,023 + 4 Ma (Kamo et al. 1996). The ~1.870 Ga peak and some ~1.8 and 2.0 – 1.87 Ga age frequency distribution plots for the northwest Yilgarn and Gascoyne provinces in Western Australia (Nelson 2008) broadly correspond to the 1.85 Ga age of the Sudbury impact structure and its ejecta layers. Peaks not known to correlate with impact events include ~3,336, ~1,625, ~1,165, ~570 and ~290 Ma.

A search for correlations between impact events and peak thermal and magmatic events is hampered by the skewed nature of the database, i.e. whereas, thousands of well dated magmatic events are known only a few large well-dated impact events have been recorded. This statistical imbalance likely reflects the difficulty in identification of impact fallout units in view of the millimeter-scale size of microkrystite spherules, corrosion of the spherules in shallow water environments, post-depositional erosion and the fact that do date only few geologists have undertaken a search for impact ejecta units.

References

Byerly GR, Lowe DR, Wooden JL, Xie X (2002) A meteorite impact layer 3470 Ma from the Pilbara and Kaapvaal Cratons. Science 297:1325–1327

Collerson KD, Kamber BS, Schoenberg R (2002) Applications of accurate, high-precision Pb isotope ratio measurement by multi-collector ICP-MS. Chem Geol 188:65–83

Condie KC (1995) Episodic ages of greenstone: a key to mantle dynamics? Geophys Res Lett 22:2215–2218

Davies GF (1995) Punctuated tectonic evolution of the Earth. Earth Planet Sci Lett 136:363–380

Glikson AY (2008) Field evidence of *Eros*-scale asteroids and impact-forcing of Precambrian geodynamic episodes Kaapvaal (South Africa) and Pilbara (Western Australia) Cratons. Earth Planet Sci Lett 267:558–570

Glikson AY, Vickers J (2007) Asteroid mega-impacts and Precambrian banded iron formations: 2.63 and 2.56 Ga impact ejecta/fallout at the base of BIF/argillite units Hamersley Basin Pilbara Craton Western Australia. Earth Planet Sci Lett 254:214–226

Glikson AY, Vickers J (2010) Asteroid impact connections of crustal evolution. Aust J Earth Sci 57:79–95

Hall HC, Bates MP (1990) The evolution of the 2.45 Ga Matachewan dyke swarm. In: Parker AJ, Rickwood PC, Tucker DH (eds) Mafic dykes and emplacement mechanisms. Balkema Rotterdam, pp 237–249

Hassler SW, Simonson BM, Sumner DY, Bodin L (2011) Paraburdoo spherule layer, Hamersley Basin, Western Australia: distal ejecta from a fourth large impact near the Archaean-Proterozoic boundary. Geology 39:307–310

Heaman LM (1997) Global mafic magmatism at 2.45 Ga: remnants of an ancient large igneous province? Geology 25:299–302

Hickman AH (2004) Two contrasting granite–greenstone terrains in the Pilbara Craton Australia: evidence for vertical and horizontal tectonic regimes prior to 2,900 Ma. Precam Res 131:153–172

Hickman AH, Van Kranendonk MJ (2004) Diapiric processes in the formation of the Archaean continental crust east Pilbara granite-greenstone terrain Australia. In: Eriksson PG, Altermann W, Nelson DR, Mueller WU, Catuneanu O (eds) The Precambrian Earth: tempos and events, Developments in Precambrian geology. Elsevier Amsterdam 12:54–75

Kamo SL, Reimold WU, Krogh TE, Colliston WP (1996) A 2.023 Ga age for the Vredefort impact event and a first report about shock metamorphosed zircons in pseudotachylitic breccias and granophyre. Earth Planet Sci Lett 144:369–388

Lowe DR, Byerly GR, Kyte FT, Shukolyukov A, Asaro F, Krull A (2003) Spherule beds 3.47–3.24 billion years old in the Barberton Greenstone Belt, South Africa: a record of large meteorite impacts and their influence on early crustal and biological evolution. Astrobiology 3:7–48

Mukasa SB, Wilson AH, Carlson RW (1998) A multi-element geochronological study of the Great Dyke, Zimbabwe: significance of the robust and reset ages. Earth Planet Sci Lett 164:353–369

Nelson DR (2008) Geochronology of the Archaean of Australia. In: De Laeter JR, Gleadow AJW, McDougall I (eds) Geochronology in Australia. Aust J Earth Sci 55:779–793

O'Reilly SY, Griffin WL, Pearson NJ, Jackson SE, Belousova EA, Alard O, Saeed A (2008) Taking the pulse of the Earth: linking crustal and mantle events. In: De Laeter JR, Gleadow AJW, McDougall I (eds) Geochronology in Australia. Aust J Earth Sci 55:983–996

Pietranik AB, Hawkesworoth CJ, Storey CD, Kemp TI, Sircombe KN, Whitehouse MJ, Bleeker RW (2008) Episodic mafic crust formation from 4.5 to 2.8 Ga: new evidence from detrital zircons Slave craton Canada. Geology 36:875–878

Poujol M, Robb LJ, Anhaeusser CR, Gericke B (2003) A review of the geochronological constraints on the evolution of the Kaapvaal Craton South Africa. Precamb Res 127:181–213

Ryder G (1990) Lunar samples lunar accretion and the early bombardment of the Moon. Eos (Trans Am Geophys Union) 71:313–322

Ryder G (1991) Accretion and bombardment in the Earth–Moon system: the Lunar record. Lunar Planet Sci Instit Contrib 746:42–43

Ryder G (1997) Coincidence in the time of the Imbrium Basin impact and Apollo 15 Kreep volcanic series: impact induced melting? Lunar Planet Sci Instit Contrib 790:61–62

Schoenberg R, Kamber B, Collerson KD, Moorbath (2003) Tungsten isotope evidence from ~3.8 Gyr metamorphosed sediments for early meteorite bombardment of the Earth. Nature 418:403–405

Simonson BM (1992) Geological evidence for an early Precambrian microtektite strewn field in the Hamersley Basin of Western Australia. Geol Soc Am Bull 104:829–839

Simonson BM, Glass BP (2004) Spherule layers—records of ancient impacts. Ann Rev Earth Planet Sci 32:329–361

Simonson BM, Hassler SW, Beukes N (1999) Late Archaean impact spherule layer in South Africa that may correlate with a layer in Western Australia. In: Dressler BO, Sharpton VL (eds) Impact cratering planetary evolution. Geol Soc Am Sp Pap 339, Boulder CO, pp 249–262

Simonson BM, Davies D, Hassler SW (2000) Discovery of a layer of probable impact melt spherules in the late Archaean Jeerinah Formation, Fortescue Group, Western Australia. Aust J Earth Sci 47:315–325

Simonson BM, Hassler SW, Beukes NJ, Sumner DY (2010) Large impacts around the Archaean-Proterozoic boundary: an update. 41st Lunar Planet Sci Conf, p 2386.pdf

Smithies RH, Van Kranendonk MJ, Champion DC (2005) It started with a plume: early Archaean basaltic proto-continental crust. Earth Planet Sci Lett 238:284–297

Van Kranendonk MJ, Smithies RH, Hickman AH, Champion DC (2007) Paleoarchaean development of a continental nucleus: the east Pilbara terrain of the Pilbara Craton Western Australia. In: Van Kranendonk MJ, Smithies R H, Bennett VC (eds) Earth's oldest rocks, developments in Precambrian geology. Elsevier Amsterdam, vol 15, pp 307–338

Chapter 11
Australian Large Impact Structures (>20 km Diameter)

Abstract The stable nature of Australian cratons allowed preservation of a comprehensive record of exposed and buried impact structures, including the classic 580 Ma Acraman-Bunyeoo impact and attended Acritarchs radiation and a number of large buried impact structures and probable impact structures, including Woodleigh, Gnargoo, Tookoonooka, Talundilly, Mount Ashmore and Warburton, identified by geophysical methods and drilling.

Not many recognize the Australian connection of Eugene Shoemaker's asteroid research legacy. Among his many achievements, together with his wife Carolyn, over a period of 12 years Eugene spent long periods exploring the Australian outback, discovering and studying impact craters, which ended with his tragic death on a Tanami Desert Road in 1997. Originally Eugene wished to become an astronaut and study the lunar craters. When this turned out impossible, he perceived Australia to be the best terrestrial terrain for studies of the scars of ancient asteroids.

The Western, central and Northern Australian crust has been remarkably stable for more than a billion years, which allowed preservation of some of the oldest impact scars and related ejecta and fallout from the collision of extra-terrestrial bodies (Tables 1.1 and 1.2). These include the oldest known impact ejecta unit (3.47 Ga, Miralga Creek, Pilbara Craton) (Sect. 8.1.1), the world's second oldest known impact structure (Yarrabubba, <2.65 Ga, Murchison region) (Sect. 11.1). A large impact structure associated with distal ejecta and mass radiation of Acritarchs (Acraman, <90 km diameter, South Australia), and the fifth largest impact structure on Earth (Woodleigh, 120 km diameter 359 ± 4 Ma, Carnarvon Basin) (Sect. 11.5). The continent includes some of the best exposed medium-scale impact craters, such as Gosses Bluff (D ~ 24 km; 142.5 Ma), Lawn Hill (D ~ 18 km), Spider (D ~ 13 km) and smaller impact craters (cf. Wolfe Creek, Hickman Crater, Henbury craters). In this section only impact structures larger than 20 km in diameter are described.

Principal features of Australian impact structures and probable impact structures larger than 20 km are described below, including (in geochronological order): Yarrabubba (<2.65 Ga; D > 50 km), Acraman (~580 Ma; D ~ 30–90 km), Shoemaker

A. Y. Glikson, *The Asteroid Impact Connection of Planetary Evolution*,
SpringerBriefs in Earth Sciences, DOI: 10.1007/978-94-007-6328-9_11,
© The Author(s) 2013

(>568 Ma; D = 29–31 km), Gnargoo (post late Permian; D = 75 km), Woodleigh (~359 Ma; D = 120 km), Warburton (pre-end Carboniferous; >200 km), Gosses Bluff (142 Ma; D = 24 km), Tookonooka and Talundilly twin impacts (115–112 Ma; 55–65 km and 84 km, respectively), Mount Ashmore (end-Eocene; >50 km).

11.1 Yarrabubba, Western Australia (D < 70 km; <2.65 Ga)

The Yarrabubba impact structure, located in the Murchison Province of the Archaean Yilgarn Craton, Western Australia, is cored by micro-granophyre surrounded by shock metamorphosed granitoids which features shatter cones, pseudotachylite veins and planar deformation features (PDF) in quartz (Fig. 11.1) (Macdonald et al. 2003). These authors suggested a diameter of 30–70 km and a minimum age of 2.56 Ga, the age of the impacted granitoid. An Ar–Ar date of 1,134 + 26 Ma is reported for pseudotachylite veins in the shocked granite (Pirajno 2005). Airborne magnetic mapping outlines a central demagnetized zone

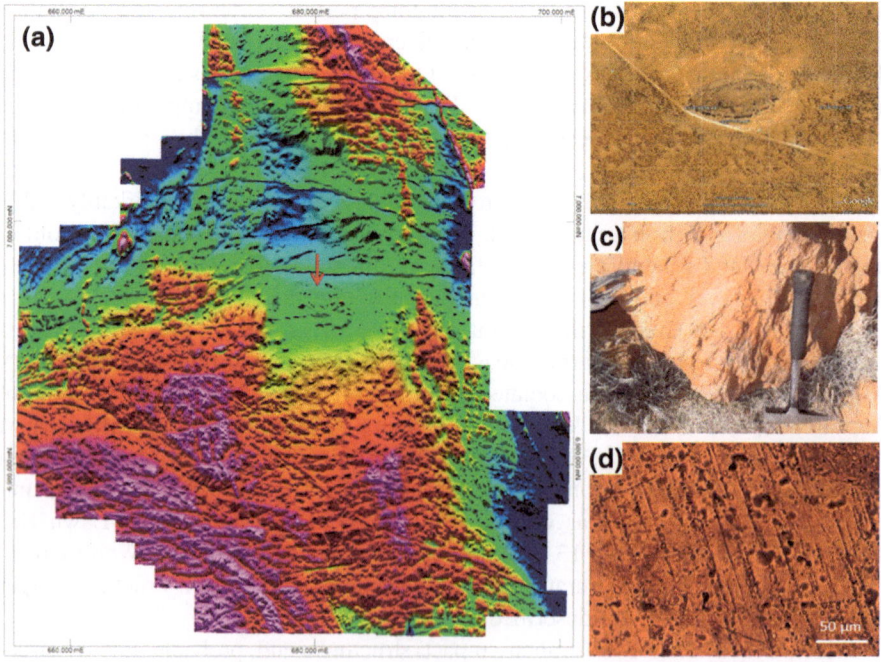

Fig. 11.1 Yarrabubba impact structure: **a** Airborne magnetic map, displaying the circular core of granophyre at the center (indicated with a *red arrow*), surrounded by semi-circular structural pattern and an eastern NNW-oriented shear zone (Courtesy Michael Jones, Impact Minerals Ltd.); **b** Landsat image of the central granophyre core (Google Earth); **c** A shatter cone in Yarrabubba Granite; **d** Planar deformation features (PDF) in quartz, Yarrabubba granite

about ~11 × 15 km large. The central granophyre plug (Barlangi Granophyre) is considered as quenched impact melt. The Yarrabubba structure is intersected by mafic dykes of likely Protoerozoic age.

11.2 Acraman, South Australia (~580 Ma; D = 30–90 km) and Bunyeroo ejecta

The discovery of the Acraman impact structure was based on identification of breccia, shatter cones, PDFs in quartz, and pseudotachylite veins in central lake outcrops. The structure forms a ~ 30 km diameter depression centered on a Lake Acraman, a ~ 20 km diameter salt lake in the Gawler Ranges, South Australia (Williams 1986; Williams et al. 1996; Williams and Gostin 2005, 2010) (Fig. 11.2). The country rocks consist of little-deformed ~1590 Ma old dacitic lavas. A close correlation was demonstrated between the Acraman impact and contemporaneous ejecta identified in the Flinders Ranges and the Officer Basin (Gostin et al. 1989, 2010; Wallace et al. 1990a, b, 1996; Hill et al. 2004). Williams and Wallace (2003) estimate the central uplift as ~40 km in diameter corresponding to the original crater, whereas an overall diameter of 85–90 km in marked by an outer fault ring.

Fig. 11.2 a Landsat scene showing most of the central part of the Acraman impact structure occupied by salt lakes and surrounded by brown outcrops of the Gawler range volcanics (NASA); **b** Outcrop of ejecta from the Acraman impact intercalated with shale of the Bunyeroo formation, flinders ranges (courtesy Victor Gostin)

Williams et al. (1996) considered a ~ 150 km diameter. K–Ar and Ar–Ar ages from pseudotachylite vein in the central uplift provide a minimum age of 450 Ma. More confident stratigraphic dating of the ejecta indicates an age of ~580 Ma (Grey et al. 2003; Williams and Wallace 2003; Hill et al. 2004).

11.3 Shoemaker, Western Australia (>568 Ma; D = 29–31 km)

The Shoemaker impact structure, originally named 'Lake Teague ring structure' (Butler 1974; Bunting et al. 1980), constitutes a circular feature superposed on the boundary between the Archaean Yilgarn Craton and the early Proterozoic Earaheedy Basin, Western Australia. These authors identified shatter cones, pseudotachylite veins and PDFs in quartz in its granite core. Detailed studies of the structure were conducted by Shoemaker and Shoemaker (1996), Pirajno et al. (2003). The structure, defined on satellite and airborne magnetic maps thanks to the incorporated banded iron formations (Hawke 2003; Pirajno et al. 2003) (Fig. 11.3), consists of a 12 km-diameter central uplift surrounded by a

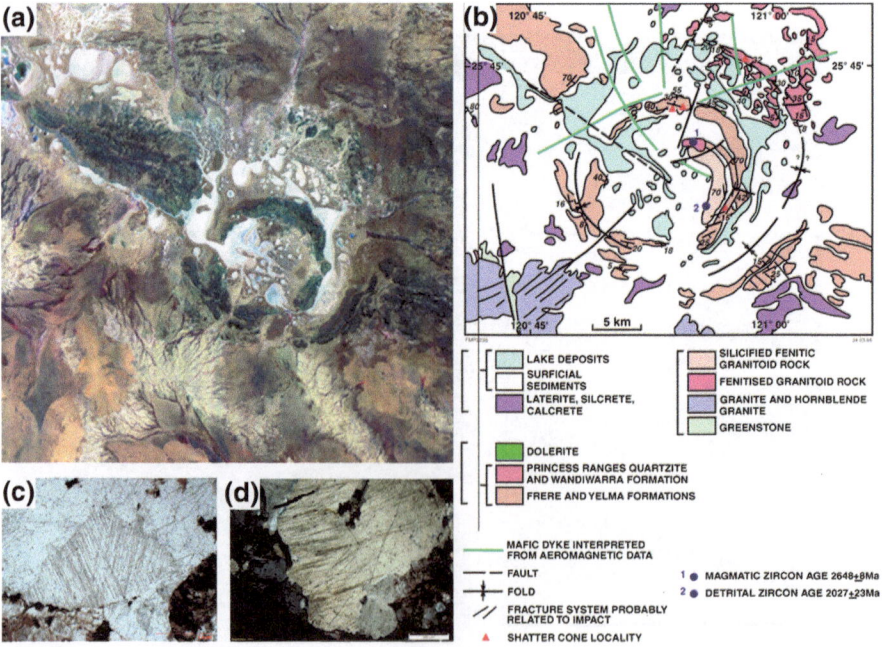

Fig. 11.3 The Shoemaker impact structure, Earaheedy basin, Western Australia: **a** Landsat image (*green*—iron oxide-dominated materials corresponding to the Frere iron formation; *red*—clay-dominated materials; **b** Geological sketch map (courtesy Franco Pirajno); **c** and **d** Planar deformation features in quartz from the silicified fenitic granitoid core (courtesy Franco Pirajno)

29–31 km-diameter ring syncline and anticline (Pirajno et al. 2003). A gravity survey defines a central negative Bouguer anomaly which coincides with the central uplift (Plescia 1999). The core consists of syenite interpreted in terms of alkali metasomatism and hydrothermal activity (Pirajno et al. 2003; Pirajno 2005). The impacted Teague Granite is dated as 2648 + 8 Ma (Pirajno et al. 2003). The minimum age of the impact is defined by the Yelma Formation which contains detrital zircons 2027 + 23 Ma (Pirajno et al. 2003). Rb–Sr granite ages of 1,630 Ma and ca 1,260 Ma (Bunting et al. 1980) reflect superposed alteration (Bunting et al. 1980; Shoemaker and Shoemaker 1996). Younger superposed K–Ar illite and smectite ages of the granite (694 + 25 Ma and 568 + 20 Ma (Pirajno et al. 2003) may reflect thermal rises associated with minor tectonic events.

11.4 Gnargoo, Western Australia (post-early Permian; D = 75 km)

The Gnargoo Structure, situated in the Southern Carnarvon Basin, is a subsurface feature identified from seismic and gravity datasets, and has been interpreted by Iasky and Glikson (2005) as a possible complex impact structure. The Gnargoo structure, buried about 500 m below Cretaceous strata, features structural elements which are remarkably identical to those of the Woodleigh impact structure, located about 275 km to the south (Iasky and Glikson 2005). These similarities include a circular Bouguer anomaly, a central structurally uplifted core comprising a buried dome with a central uplifted plug, a weakly defined inner 10 km-diameter circular Bouguer anomaly surrounded by a broadly circular zone ~75 km in diameter. The outer ring intersects the N–S Bouguer anomaly lineament of the Giralia Range. The structure comprises a layered sedimentary dome of Ordovician to Lower Permian strata, surrounding a cone-shaped, central uplift plug of 7–10 km diameter. Seismic-reflection data indicate a minimum central structural uplift of 1.5 km. An interpretation of Gnargoo as a salt dome is unlikely as no extensive evaporite units are known in the Southern Carnarvon Basin. Morphometric estimates of the rim-to-rim diameter based on seismic data for the central dome correspond to the observed diameter deduced from gravity data, and fall within the range of morphometric parameters of known impact structures. The age of Gnargoo is constrained between the deformed Lower Permian target rocks and unconformably overlying Lower Cretaceous strata.

11.5 Woodleigh, Western Australia (~359 Ma; D = 120 km)

Woodleigh is a large buried impact structure situated in the Gascoyne Platform of the Southern Carnarvon Basin, Western Australia, and originally identified from a distinctive multi-ringed gravity signature (Iasky et al. 1998; Iasky and Mory 1999; Mory et al. 2000; Glikson et al. 2005a, b). Woodleigh-1 hole penetrated shock-metamorphosed granitoid basement (171–333 m) beneath undisturbed Jurassic sediments. An impact origin of Woodleigh structure is clearly demonstrated by the presence of PDFs in quartz, indicating shock pressures of at least 20 GPa, and other shock-induced phases in the granitoid interesected by the Woodleigh-1 drill hole 1 (Mory et al. 2000a, b; Hough et al. 2003; Reimold et al. 2003; Whitehead et al. 2003; Glikson et al. 2005a). The central granitoid displays evidence of considerable clay alteration interpreted as an impact-induced hydrothermal effect (Glikson et al. 2005a; Pirajno 2005). Enrichment of Mg and siderophile elements in the shocked gneiss may reflect contamination by the projectile and/or hydrothermal fluids (Glikson et al. 2005a). Iasky and Mory (1999), Mory et al. (2000a, b, 2001), Iasky et al. (2001), Glikson et al. (2005b) have inferred a diameter of ~120 km based mainly on geophysical arguments and morphometric analysis. In contrast, Reimold and Koeberl (2000) suggested that the structure may be only 40 km in diameter, while Reimold et al. (2003) gave a somewhat higher estimate of ~60 km. Based on shock-pressure attenuation down hole, Whitehead et al. (2003) estimated ~6 km of uplift in Woodleigh 1, which would equate to a crater size between the suggested extremes. Hough et al. (2003) discussed the various size constraints and concluded that the original crater was probably nearer to 60 km than 120 km in diameter. Stratigraphic constraints allow an age between Devonian and Early Jurassic (Iasky and Mory 1999; Mory et al. 2000a). Uysal et al. (2001) reported an age of 359 + 4 Ma, within error of the Devonian— Carboniferous boundary (recently refined to 359.2 + 2.5 Ma: Gradstein et al. 2004), based on K–Ar dating of fine-grained (52 mm) illitic clay minerals, interpreted as impact-induced hydrothermal alteration products. In contrast, Renne et al. (2002) questioned the inferred link between clay mineral paragenesis and impact-related features, and stated their opinion that the impact could have been much older than mid-Devonian. A recent seismic transect by Geoscience Australia indicates a diameter of 51 km at deep crustal level of Woodleigh, likely representing the base of the impact-deformed zone.

11.6 Warburton Probable Impact Structure, North-East South Australia

The East Warburton Basin, northeast South Australia (Glikson et al. 2013), features major geophysical anomalies, including a magnetic high of near-200 nT centered on a ~25 km-wide magnetic low (<100 nT), corresponding to density estimates of 2.8–2.99 g/cm^3 and interpreted in terms of a homogeneous magmatic body below 6 km depth (Meixner et al. 2000). A distinct seismic tomographic low velocity anomaly associated with the East Warburton Basin may reflect a thick (~9.5 km) sedimentary section, high temperatures and possible deep fracturing (Saygin and Kennett 2010, 2012). The anomaly is similar to a tomographic anomaly associated with the 120 km-diameter Woodleigh impact structure, Western Australia (Figs. 5.1, 5.2) (Saygin and Kennett 2012). Scanning electron microscope (SEM) analyses of the intrusive Big Lake Suite granites resolve microbreccia veins of micron-scale particles injected into resorbed quartz grains. Planar to sub-planar microstructures in quartz grains are documented in granite, volcanic and sedimentary cores from 55 drill holes within the >30,000 km^2-large East Warburton Basin. Quartz grains may display multiple intersecting planar and sub-planar elements and include relic lamella less than 2 micron wide with inter-planar spacing of 4–5 μ. Planar to sub-planar elements are deformed, displaying bent and wavy patterns commonly accompanied with fluid inclusions. Universal stage measurements (total of 243 planar and sub-planar sets in 157 quartz grains) indicate dominance of Miller indices diagnostic of shock metamorphism ($\prod\{10\text{–}12\}$, $\omega\{10\text{–}13\}$ and subsidiary $\S\{11\text{–}22\}$, $\{22\text{–}41\}$, $r\{10\text{–}11\}$ and $x\{51\text{–}61\}$). Transmission Electron Microscopy (TEM) analysis displays relic narrow ≤ 1 μm-wide lamellae and relic non-sub grain boundaries where crystal segments maintain optical continuity, distinguishing the lamellae from tectonic and low shock-pressure basal 'Boehm lamellae' $\{0001\}$. Extensive sericite alteration of feldspar suggests hydrothermal alteration to a depth of ~500 m below the top unconformity which truncates the quartz microstructure-bearing East Warburton Basin terrain. A complete gradation occurs between relic Qz/PDF and deformed Qz/PDF such as occur in several large impact structures. An uplift of the Big Lake Granite Suite by 4–5 km during ca 298–295 Ma, followed by erosion, suggested by the lack of upper Ordovician to Devonian strata in the East Warburton Basin, indicates major tectonism in the end-Carboniferous. The definition of an impact structure in the East Warburton Basin requires corroboration by deep crustal seismic reflection transects. Evidence for the origin of quartz lamellae in the East Warburton Basin includes: (a) Agreement of measured $C_{OAQZ}{}^\wedge P_{PE}$ angles with Miller indices, mainly $\omega\{10\text{–}13\}$ and $\prod\{10\text{–}12\}$, indicative of shock levels higher than 20 GPa (French 1998; Langenhorst 2002); (b) Observation by optical microscopy and TEM of lamellae <1–2 μm wide with ~4–5 Nm spacing; (c) The occurrence of multiple intersecting lamellae, and (d) TEM observation of relic non-sub-grain boundaries between planar segments in optically coherent host quartz. The planar elements are associated with, or overprinted by, wavy and bent wide and wide-spaced planar features which form low angle boundaries between subgrains of divergent low angle optic orientation. Re-deformation, recrystallization and annealing of quartz lamellae, which contain relic features corresponding to Qz/PDF, are likely to be associated with post shock centripetal-oriented deformation inherent in the formation of a central uplift as well as hydrothermal activity triggered by impact (Naumov 2002; Pirajno 2005). The Warburton Basin sediments and the Big Lake Suite granite have experienced significant multi-deformation phases and associated hydrothermal events in the Early Permian, Permo-Triassic, Mid-Cretaceous and Tertiary (McLaren and Dunlap 2006). North and west of the East Warburton Basin and separated by a structural ridge (Birdsville Track Ridge) is the West Warburton basin, also known as the Poolowanna Basin. The Poolowanna Basin is marked by a distinct magnetic anomaly and by a strong low velocity tomography anomaly which represents a short seismic period of up to 8.3 s, similar to the East Warburton Basin (Fig. 11.4) (Saygin and Kennett 2010,

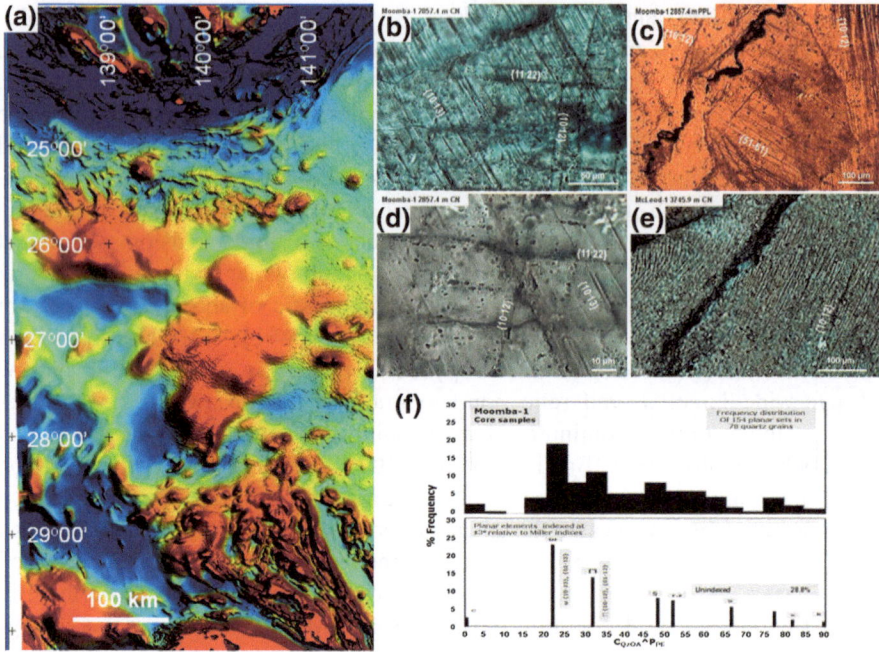

Fig. 11.4 The Warburton probable impact structure: **a** Airborne magnetic map, shock metamorphism is observed at and around the central magnetic high; **b–e** Planar deformation features (PDF) in quartz grains from drill holes in end Carboniferous granite of the Warburton structure; **f** Frequency distribution of C axis to poles to PDF angles, indicating prevalence of Miller indices {10–12}, {10–13} indicative of shock metamorphism at levels >10 GPa

2012). Somewhat weaker anomalies pertain to the 12.5 s period. However, only a relatively thin sedimentary thickness pertains to the Poolowanna basin, hinting at yet unresolved factors. The age of the Warburton impact precedes the age of the youngest shock metamorphosed Innamincka granite of ~289 Ma and could potentially correspond to the end-Devonian impact cluster.

11.7 Gosses Bluff, Northern Territory (142 Ma; D = 24 km)

Dietz (1967) identified shatter cones in the 24 km-diameter Gosses Bluff ring structure, central Australia (Figs. 1.1, 4.1) (Crook and Cook 1966; Glikson 1969; Brett et al. 1970; Milton et al. 1972, 1996a, b). The well-exposed core of the central uplift forms a ~5 km-diameter cliff ring surrounding a shale-dominated central depression. The target rocks are flat-lying Palaeozoic sediments of the Amadeus Basin, predominantly sandstone and shale. The impact was considered by Crook and Cook (1966), suggesting alternative impact origin or cryptovolcanic origin. A joint Bureau of Mineral Resources and US Geological Survey study in the late 1960s (Glikson 1969; Milton et al. 1972, 1996a, b; Milton 1972) confirmed an impact origin from oriented shatter cones, PDFs in quartz, impact melt rocks, and evidence for a centripetal sense of deformation. The structure is very well defined by seismic evidence and gravity data, including a ~3 km diameter annular gravity low which coincides with the central uplift and gravity data defining the outer boundary at ~24 km (Barlow 1990; Milton et al. 1996a). Relics of melt breccia are found at Mt Pyroclast and in drill holes, which comprise melt breccia, partly melted fragments of recrystallised sandstone, baked mudstone, and devitrified flow banded silica glass. $^{39}Ar/^{40}Ar$ dating of melt rock yielded an age of 142.5 + 0.8 Ma (Milton and Sutter 1987), within error from the Jurassic–Cretaceous boundary (145.5 + 4.0 Ma: Gradstein et al. 2004).

11.8 Tookoonooka (~125 Ma; D = 55–65 km) and Talundilly (D = 84 km; ~125 Ma) Twin Impact Structures, Eromanga Basin, Queensland

The 55–65 km-large Tookoonooka impact structure (Gorter et al. 1989; Gostin and Therriault 1997; Bron and Gostin 2012), buried below ~900 m thick Cretaceous and Tertiary flat-lying strata of the Eromanga Basin, is expressed on seismic sections as a concentric arrangement of anticlines and synclines surrounding a 22 km-wide complex central dome associated with central gravity low and magnetic high. The inferred crater diameter is given as ~55 km by Gorter et al. (1989) and Gorter (1998), but as ~66 km by Gostin and Therriault (1997). Regional well intersections over an area of 400,000 km^2 intersect Lower Cretaceous strata (basal Wyandra Sandstone, Cadna Owie Formation) containing impact melt breccia which includes quartz grains with PDF corresponding to Miller indices {10–12}, {11–22}, {21–31} and {11–21}, confirming an impact origin for Tookoonooka (Gostin and Therriault 1997; Bron and Gostin 2012). The clasts contain accretionary and melt components and lithic and mineral grains which correspond to the Tookoonooka target rock sequence, including basement. The timing of the impact event is confirmed to be the Barremian-Aptian boundary, at 125 ± 1 Ma. Impact occurred during sedimentation of the Cadna-Owie Formation, dated by Gorter et al. 1989 and Gostin and Therriault 1997 as 128 + 5 Ma.

The Talundilly Structure (Longley 1989; Gorter and Glikson 2012), located about 328 km northeast of Tookoonooka, is represented on seismic reflection transects as a major seismic anomaly about 84 km in diameter, disrupting the early Cretaceous Wyandra Sandstone, located between the Cadna Owie and Bulldog Shale formations. The seismic anomalous zone coincides with a prominent aeromagnetic (TMI) high centrally located within a near-circular seismic anomaly. The structure consists of a raised central area, with radial faults extending from the central high, an annular synform with disrupted seismic elements dipping at low angles towards the central uplift, and an outer faulted rim. Talundilly-1 well located about 30 km northwest of the structural peak intersected arenite which contains quartz grains with PDF. The age of the structure, as determined from seismic correlation and sparse palynology, is estimated as ~125 Ma, coinciding with the age of Tookoonooka and a marine transgression.

11.9 Mount Ashmore Probable Impact Structure, Timor Sea

The Mount Ashmore dome, west Bonaparte Basin, Timor Sea, is located below a major pre-Oligocene post-Late Eocene unconformity and above a ~6 km-deep-seated basement high indicated by marked gravity and magnetic anomalies (Glikson et al. 2010) (Fig. 5.3). Whereas, the lower diameter of the dome is approximately 50 km and although the rim syncline is not well defined, the overall structure is estimated as larger than 100 km in diameter. The core of the dome features chaotic deformation and centripetal kinematic deformation patterns. Well intersections reveal micro-brecciation, comminuted and flow-textured fluidization of altered sedimentary material. The microbreccia is dominated by aggregates of poorly diffracting micrometer to tens of micrometers-scale to sub-millimeter particles, including relic sub-planar fractured quartz grains, carbonate, barite, apatite and K-feldspar. No volcanic material or evaporates were encountered, militating against interpretations of the structure in terms of magmatic intrusion or salt diapirism, models which are also inconsistent with strong gravity and magnetic anomalies reflecting a basement high below the dome. Whereas an impact origin cannot be proven due to the lack of shock metamorphic effects, such as PDF, impact melt or coesite, an impact model is consistent with the chaotic structure of the dome-structured core, centripetal sense of deformation, micro-brecciation and fluidization of the Triassic to Eocene rocks. An analogy can be drawn between the Mt Ashmore structural dome and impact structures formed in volatile-rich sediments where shock is attenuated by high volatile pressure. The Mount Ashmore dome is contemporaneous with a Late Eocene impact cluster (Popigai: D ~ 100 km, 35.7 + 0.2 Ma; Chesapeake Bay: D ~ 85 km, 35.3 + 0.1 Ma) (Table 12.1).

References

Barlow NG (1990) Estimating the terrestrial crater production rate during the late heavy bombardment period. Lunar Planet Instit Contrib 746:4–7

Brett R, Guppy DJ, Milton DJ (1970) Two circular structures of impact origin in Northern territory Australia. Meteorit 5:184

Bron KA, Gostin V (2012) The Tookoonooka marine impact horizon Australia: Sedimentary and petrologic evidence. Meteor Planet Sci 47:296–318

Bunting JA, de Laeter JR, Libby WG (1980) Evidence for the age and cryptoexplosive origin of the Teague ring structure Western Australia. Geol Surv W Aust Ann Rev 1980:81–85

Butler H (1974) The Lake Teague ring structure Western Australia: an astrobleme? Search 5:536–537

Crook KA, Cook PJ (1966) Gosses Bluff diaper, cryptovolcanic structure, or astroblemes. J Geol Soc Aust 13:495–516

Dietz RS (1967) Shatter cone orientation at Gosses Bluff astrobleme. Nat 216:1082–1084

French BM (1998) Traces of catastrophe—a handbook of shock metamorphic effects in terrestrial meteorite impact structures, 120 pp. Lunar Planet Sci Instit Contrib 954

Glikson AY (1969) Geology of the outer zone of the Gosses Bluff crypto-explosion structure, Northern Territory. Bur Min Resour Record 1969/42

Glikson AY, Eggins S, Golding S, Haines P, Iasky RP, Mernagh TP, Mory AJ, Pirajno F, Uysal IT (2005a) Microchemistry and microstructures of hydrothermally altered shock-metamorphosed basement gneiss, Woodleigh impact structure, southern Carnarvon basin, Western Australia. Aust J Earth Sci 52:555–573

Glikson A, Mory AJ, Iasky R, Pirajno F, Golding S, Uysal IT (2005b) Woodleigh, Southern Carnarvon basin, Western Australia: history of discovery late Devonian age and geophysical and morphometric evidence for a 120 km-diameter impact structure. Aust J Earth Sci 52:545–553

Glikson AY, Jablonski D, Westlake S (2010) Origin of the Mount Ashmore structural dome west Bonaparte basin Timor Sea. Aust J Earth Sci 57:411–430

Glikson AY, Uysal IT, Fitz Gerald JD, Saygin E (2013) Geophysical anomalies and quartz microstructures, Eastern Warburton Basin, North-east South Australia: tectonic or impact shock metamorphic origin? Tectonophysics 589:57–76

Gorter JD (1998) The petroleum potential of Australian Phanerozoic impact structures. Aust Petrol Explor J 37:159–186

Gorter JD, Glikson AY (2012) Talundilly Western Queensland Australia: geophysical and petrological evidence for an 84 km-large impact structure and an early cretaceous impact cluster. Aust J Earth Sci 59:51–73

Gorter JD, Gostin VA, Plummer P (1989) The Tookoonooka Structure: an enigmatic sub-surface feature in the Eromanga basin its impact origin and implications for petroleum exploration In: O'Neil BJ (ed) The Cooper and Eromanga basins Australia: Proceedings of the Cooper and Eromanga Basins conference Adelaide, pp 441–456

Gostin VA, Therriault AM (1997) Tookoonooka: a large buried early Cretaceous impact structure in the Eromanga basin of southwestern Queensland Australia. Meteor Planet Sci 32:593–599

Gostin VA, Keays RR, Wallace MW (1989) Iridium anomaly from the Acraman ejecta horizon: impacts can produce sedimentary iridium peaks. Nat 340:542–544

Gostin VA, McKirdy DM, Webster LJ, Williams GE (2010) Ediacaran ice-rafting and coeval asteroid impact, South Australia: Insights into the terminal Proterozoic environment. Aust J Earth Sci 57(7):859–869

Gradstein FM, Ogg JG, Smith AG, Bleeker W, Laurens LJ (2004) A new geologic timescale with special reference to Precambrian and Neogene. Episodes 72:83–100

Grey K, Walter MR, Calver CR (2003) Neoproterozoic biotic diversification: snowball earth or aftermath of the Acraman impact? Geol 5:459–462

Hawke PJ (2003) Interpretation of geophysical data over the Shoemaker impact structure Earaheedy basin Western Australia. Geol Surv West Austr Record 2003/6

Hill AC, Grey K, Gostin VA, Webster LJ (2004) New records of late Neoproterozoic Acraman ejecta in the officer basin. Aust J Earth Scis 38:291–298

Hough R, Lee MR, Bevan AWR (2003) Characterization and significance of shocked quartz from the Woodleigh impact structure, Western Australia. Meteor Planet Sci 38:1341–1350

Iasky RP, Glikson AY (2005) Gnargoo: a possible 75 km-diameter post-early Permian—pre-Cretaceous buried impact structure Carnarvon basin Western Australia. Aust J Earth Sci 52:577–586

Iasky RP, Mory AJ (1999) Geology and petroleum potential of the Gascoyne platform Southern Carnarvon basin. Western Australia Geol Surv West Aust, Report 69

Iasky RP, Mory AJ, Shevchenko SI (1998) A structural interpretation of the Gascoyne platform Southern Carnarvon basin West Australia In: Purcell PG, Purcell RR (eds) The sedimentary basins of West Australia, Proc Petrol Explor Soc Aust Symp Perth WA pp 589–598

Iasky RP, Mory AJ, Blundell KA (2001) The geophysical signature of the Woodleigh impact structure Southern Carnarvon basin, Western Australia. Geol Surv West Aust Report 79

Langenhorst F (2002) Shock metamorphism of some minerals: basic introduction and micro-structural observations. Bull Czech Geol Surv 77(4):265–282

Longley IM (1989) The Talundilly anomaly and its implications for hydrocarbon exploration of Eromanga astroblemes In: O'Neil BJ (ed) The Cooper and Eromanga basins Australia, Proceedings of the Cooper and Eromanga basins conference, Adelaide. 473–490

Macdonald FA, Bunting JA, Cina SE (2003) Yarrabubba—a large deeply eroded impact structure in the Yilgarn Craton Western Australia. Earth Planet Sci Lett 213:235–247

McLaren S, Dunlap WJ (2006) Use of 40Ar/39Ar K-feldspar thermo-chronology in basin ther-mal history reconstruction: an example from the big lake Suite granites Warburton basin. South Aust Basin Res 18:189–203

Meixner TJ, Gunn PJ, Boucher RK, Yeats AN, Murra, L, Yeats TN, Richardson LM, Freares RA (2000) The nature of the basement to the cooper basin region. South Aus Expl Geophys 31:024–032

Milton DJ (1972) Structural geology of the Henbury meteorite craters Northern Territory Australia. US Geol Surv Profess Pap 599:C1–C17

Milton DJ, Barlow BC, Brett R, Brown AR, Manwaring EA, Moss FJ, Sedmik ECE, Van Son J, Young GA (1972) Gosses bluff impact structure australia. Science 175:1199–1207

Milton DJ, Sutter JF (1987) Revised age for the Gosses Bluff impact structure Northern territory Australia based on 40Ar/39Ar dating. Meteorit 22:281–289

Milton DJ, Glikson AY, Brett R (1996a) Gosses Bluff—a latest Jurassic impact structure cen-tral Australia: Part 1: geological structure stratigraphy and origin. Aust Geol Surv Org J Aust Geol Geophys 16:453–486

Milton DJ, Barlow BC, Brown AR, Moss FJ, Manwaring EA, Sedmik ECE, Young GA, Van Son J (1996b) Gosses Bluff—a latest Jurassic impact structure central Australia: Part 2- seismic magmatic and gravity studies. Aust Geol Surv Org J Aust Geol Geophys 16:487–527

Mory AJ, Iasky RP, Glikson AY, Pirajno F (2000a) Woodleigh Carnarvon basin, Western Australia: a new 120 km-diameter impact structure. Earth Planet Sci Lett 177:119–128

Mory AJ, Iasky RP, Glikson AY, Pirajno F (2000b) Response to 'Critical comment on AJ Mory et al. (2000) Woodleigh Carnarvon basin Western Australia: a new 120 km diameter impact structure. Earth Planet Sci Lett 184:359–365

Mory AJ, Pirajno F, Glikson AY, Crocker A (2001) Geol Surv West Aust Woodleigh 1, 2, 2A well completion reports, Gascoyne platform, Southern Carnarvon Basin. Western Australia Geol Surv of West Aust Record (2001/6)

Naumov MV (2002) Impact generated hydrothermal systems. In: Plado J, Pesonen LJ (eds) Impacts in Precambrian Shields, Springer, Berlin, pp 117–173

Pirajno F (2005) Hydrothermal processes associated with meteorite impact structures: evidence from three Australian examples and implications for economic resources. Aust J Earth Sci 52:587–606

Pirajno F, Hawke P, Glikson AY, Haines PW, Uysal T (2003) Shoemaker impact structure Western Australia. Aust J Earth Sci 50:775

Plescia JB (1999) Gravity signature of Teague ring impact structure Western Australia. Geol Soc Am Sp Pap 339:165–175

Reimold WU, Koeberl C (2000) Critical comment on: AJ Mory et al 'Woodleigh Carnarvon basin Western Australia: a new 120 km diameter impact structure. Earth Planet Sci Lett 184:353–357

Reimold WU, Koeberl C, Hough RM, Mcdonald I, Bevan A, Amare K, French BM (2003) Woodleigh impact structure Australia: shock petrography and geochemical studies. Meteor Planet Sci 38:1109–1130

Renne PR, Reimold WU, Koeberl C, Hough R, Clayes P (2002) Comment on 'K–Ar evidence from illitic clays of a late Devonian age for the 120 km diameter Woodleigh impact structure Southern Carnarvon basin Western Australia. Earth Planet Sci Lett 201:247–252

Saygin E, Kennett BLN (2010) Ancient seismic tomography of Australian continent. Tectonophysics 481:116–125

Saygin E, Kennett BLN (2012) Crustal structure of Australia from ambient seismic noise tomography. J Geophys Res 117:B01304

Shoemaker EM, Shoemaker CS (1996) The Proterozoic impact record of Australia. Aust Geol Surv Org J Aust Geol Geophys 16:379–398

Uysal IT, Golding SD, Glikson AY, Mory AJ, Glikson M (2001) K–Ar evidence from illitic clays of a late Devonian age for the 120 km diameter Woodleigh impact structure Southern Carnarvon basin Western Australia. Earth Planet Sci Lett 192:281–289

Wallace MW, Gostin VA, Keays RR (1990a) Acraman impact ejecta and host shales—evidence for low-temperature mobilization of iridium and other platinoids. Geol 18:132–135

Wallace MW, Gostin VA, Keays RR (1990b) Spherules and shard-like clasts from the late Proterozoic Acraman impact ejecta horizon South Australia. Meteorit 25:161–165

Wallace MW, Gostin VA, Keays RR (1996) Sedimentology of the Neoproterozoic Acraman impact-ejecta horizon South Australia. Aust Geol Surv Org J Aust Geol Geophys 16:443–451

Whitehead AD, Grieve RAF, Spray J (2003) Planar deformation features in the Woodleigh impact structure Western Australia and their bearing on the degree of structural uplift in the structure. Geol Soc Am, NE Section, 38th Ann Meet Pap. pp 7–18

Williams GE (1986) The Acraman impact structure; source of ejecta in late Precambrian shales, South Australia. Science 233:200–203

Williams GE, Gostin VA (2005) The Acraman—Bunyeroo impact event (Ediacaran) South Australia and environmental consequences: 25 years on. Aust J Earth Sci 52:607–620

Williams GE, Gostin VA (2010) Geomorphology of the Acraman impact structure, Gawler Ranges, South Australia. Cadernos Laboratorio Xeoloxico de Laxe 35:209–220

Williams GE, Wallace MW (2003) The Acraman asteroid impact South Australia: magnitude and implications for the late Vendian environment. J Geol Soc London 160:545–554

Williams GE, Schmidt PW, Boyd DM (1996) Magnetic signature and morphology of the Acraman impact structure South Australia. Aust Geol Surv Org J Aust Geol Geophys 16:431–442

Chapter 12
Impacts and Mass Extinctions

Abstract Since establishment of the connection between the Cretaceous-Tertiary (K-T) impact event and attendant mass extinction 65 Ma-ago, another established connection is demonstrated at 580 Ma where the Acraman-Bunyeroo impact (South Australia) coincides with the radiation of Acritarchs. Possible impact-extinction relations also occur at the end-Devonian, where the age of Woodleigh (120 km-large) coincides with the demise of coral reefs.

When a large (>200 m) asteroid hits the solid surface at a high angle it penetrates to a depth approximately 1.5 times its diameter, depending on the rheology of the impacted rocks, where its kinetic energy is translated into heat, triggering an explosion, fragmentation, cratering, melting and vaporization of the immediately surrounding rocks. In craters larger than about 4 km the Earth's crust rebounds to form a central uplift (French 1998) (Fig. 4.2). Depending on the size of the impact, seismic waves propagate, leading to earthquakes, faulting and tsunami waves over large regions. Environmental effects of asteroid impacts include the initial fireball flash, mega-tsunami waves, release of aerosols (dust, sulfur dioxide, carbon soot), acid rain and release of greenhouse gases (water, CO_2, methane, nitrous oxide) from cratered regions, leading to ocean acidification. This leads to an *asteroid winter* phenomenon, with some 10–20 % of solar radiation blocked for 8–13 years (Pope et al. 2004), followed by a greenhouse-gas induced warm period lasting centuries to millennia. Species which have escaped the immediate regional and transient effects of large impacts were affected by the long-lasting consequences of the well-mixed greenhouse gases, mainly CO_2, Nitrous oxide and methane. CO_2 stays in the atmosphere for thousands to tens of thousands years, leading to extended periods of high global temperatures, compounding the effects on the biosphere. Effects on the oceans include acidification, oxygen depletion and consequent anoxic conditions and toxic H_2S forming emanations (Ward 1994, 2007, 2009).

Repeatedly impacts by asteroid clusters and major volcanic episodes induced abrupt changes in the composition of the atmosphere and ocean, triggering mass extinction of species. The impacts spread global dust clouds, deposited as thin iridium-rich layers. Unique among the elements, iridium is enriched in asteroids to several hundred parts per billion, more than two orders of magnitude higher than

A. Y. Glikson, *The Asteroid Impact Connection of Planetary Evolution*,
SpringerBriefs in Earth Sciences, DOI: 10.1007/978-94-007-6328-9_12,
© The Author(s) 2013

in the Earth crust and mantle, which allows the tracing of extraterrestrial chemical signatures in dispersed fallout material. The discovery of the KT boundary impact and its coincidence with the end- Cretaceous mass extinction ~65 Ma-ago (Alvarez et al. 1980; Alvarez 1986) heralded a major shift in the long-running debate regarding the respective role of uniformitarian principles vs catastrophic events in natural evolution (Stanley 1987; Raup 1991; Sepkoski 1996; Watson 2008). A large body of stratigraphic and isotopic age evidence points to close temporal overlaps, within age determination error, between extraterrestrial impacts, volcanic events and mass extinction of species in the Phanerozoic (Glikson 2005, 2008; Keller 2005) (Table 12.1; Fig. 1.2). Whereas in itself an age coincidence does not prove cause-and-effect relationship between impact, volcanism and mass extinction, large impacts and major volcanic eruptions inevitably had major atmospheric effects and thus a role in contemporaneous mass extinction events. The observed overall directionality of complexity, brain development and intelligence of species across extinction boundaries (Watson 2008) may be attributable to genetic transmission by the surviving species (Convey Morris 2003).

The evidence for a connection between asteroid impact and mass extinction at the KT boundary is conclusive. In March, 2011, 41 international experts from 33 institutions concluded it was the *Chicxulub* impact (Chicxulub, 64.98 ± 0.05; D = ~150–170 km; Hildebrand et al. 1991) which triggered mass extinctions at the KT boundary, including those of dinosaurs. This was established from extensive analysis of evidence from paleontology, geochemistry, climate modeling, geophysics and sedimentology. A contemporaneous impact is represented by Boltysh (65.17 ± 0.64 Ma; D = 24 km) (Boltysh, Table 1.1). Whereas only a single ejecta unit exists at the KT boundary, the existence of other synchronous impacts representing fragments of the same asteroid cannot be ruled out and would be consistent with observation of impact clusters elsewhere.

Most large vertebrates on land, sea and air, including *dinosaurs*, *plesiosaurs*, *mosasaurs*, and *pterosaurs*, and most plankton and tropical invertebrates, including reef-dwellers, became extinct at the KT boundary. Many land plants were severely affected. However, insects, mammals, birds, and flowering plants on land, as well as fish, corals, and molluscs in the ocean, survived and diversified. Apart from the KT mass extinction, events associated with extraterrestrial impact and volcanism included (Fig. 1.2; Table 1.2, Keller 2005, and a summary in Table 12.1):

1. Late to end-Devonian (~364 Ma and ~359 Ma, respectively), representing a period of intense asteroid bombardment, when ~19 % of all families and 50 % of all genera became extinct (Benton 2003). Groups most affected included *trilobites*, *stromatoporoids*, *ammonoids*, *rugosecorals*, *placoderms*, *cricoconaids*, *agnathams* and other.
2. The most extreme mass extinction event occurred at the Permian-Triassic boundary (251 + 0.4 Ma), a period of intense volcanic activity (Siberian Norilsk 251.7 + 0.4 − 251.1 + 0.3 Ma) and at least one large asteroid impact with an overlapping age (*Araguinha*, Brazil—252.7 ± 3.8 Ma; 40 km, affecting carbonates and shale target rocks releasing CO_2). At this stage ~57 % of

Table 12.1 Phanerozoic stage boundaries, large asteroid impacts and correlated volcanic and tectonic events

Stage boundaries/epochs	Large asteroid impacts	Large volcanic provinces	Percentage mass extinction of genera (%)
Mid-Miocene Langhian 15.97 Ma	Ries (24 km) 15.1 ± 1.0 Ma	Columbia plateau basalt 16.2 ± 1 Ma	6
Eocene–Oligocene boundary 33.9 ± 0.1 Ma	Popigai (100 km) 35.7 ± 0.2 Ma; Chesapeake Bay (85 km) 35.5 ± 0.3 Ma Mount Ashmore: E-O Boundary	Ethiopian basalts 36.9 ± 0.9 Ma	10
KT boundary 65.5 ± 0.3 Ma	Chicxulub (170 km) 64.98 ± 0.05 Ma Boltysh (25 km) 65.17 ± 0.64 Ma	Deccan plateau basalts. 65.5 ± 0.7 Ma (pooled Ar ages: 65.5 ± 2.5 Ma)	46
Cenomanian–Turonian 93.5 ± 0.8 Ma	Steen river (25 km) 95 ± 7 Ma	Madagascar basalts 94.5 ± 1.2 Ma	17
Aptian (early Cretaceous) 125 – 112 Ma	Carlswell (39 km) 115 ± 10 Ma; Tookoonooka (55 km; 125 ± 1 Ma); Talundilly (84 km; 125 ± 1 Ma; Mien (9 km) 121 ± 2.3 Ma; Rotmistrovka (2.7 km) 120 ± 10 Ma	Ontong-Java LIP 120 Ma Kerguelen LIP 120 – 112.7 – 108.6 Ma Ramjalal Basalts, 117 ± 1	14
End-Jurassic 145.5 ± 4 Ma	Morokweng (70 km) 145 ± 0.8 Gosses Bluff (24 km) 142.5 ± 0.8 Ma; Mjolnir (40 km) 143 ± 2.6 Ma	Dykes SW India 144 ± 6 Ma	20
End-Pliensbachian 183 ± 1.5 Ma		Peak Karoo volcanism start 190 ± 5 Ma; peaks 193, 178 Ma; Lesotho 182 ± 2 Ma	19
End-Triassic 199.6 ± 0.3 Ma		Central Atlantic igneous province: 203 ± 0.7 to 199 ± 2 Ma Newark Basalts 201 ± 1 Ma	18
Norian/Rhatian 216.5	Manicouagan (100 km) 214 ± 1 Ma; Rochechouart (23 km) 213 ± 8 Ma		34
Permian–Triassic: 251 ± 0.4 Ma; 251.4 ± 0.3 to 250.7 ± 0.3 Ma	Araguinha (40 km) 252.7 ± 3.8 Ma	Siberian Norilsk 251.7 ± 0.4 to 251.1 ± 0.3 Ma	80

(continued)

Table 12.1 (continued)

Stage boundaries/epochs	Large asteroid impacts	Large volcanic provinces	Percentage mass extinction of genera (%)
Late to end Devonian 374 – 359 Ma	Woodleigh (120 km) 359 ± 4 Ma; Siljan (52 km) 361 ± 1.1 Ma; Alamo breccia (~100 km) ~ 360 Ma; Charlevoix (54 km) 342 ± 15 Ma	Rifting and 364 Ma Pripyat–Dneiper–Donets volcanism	30, 58
End-ordovician 443.7 ± 1.5 Ma	Several small poorly dated impact craters		60
End-early Cambrian 513 ± 2 Ma	Kalkarindji volcanic Province, northern Australia 507 ± 4 Ma		42

Note No genetic connection is necessarily implied in this paper between age-correlated events

families, ~83 % of genera, ~95 % of marine species and over 70 % of terrestrial species became extinct (Keller 2005) (Fig. 1.2). Groups most affected were *echinoderms, bryozoans*, small and complex *foraminifera, rugose corals, calcic sponges, gastropods, bivalves* and *radiolarian*.

3. End-Triassic (214 – 201 Ma) opening of the Atlantic Ocean, volcanism and asteroid bombardment resulted in the extinction of ~23 % of families and ~48 % of all genera, including *ammonoids* and *conodonts*.

4. Approximately 1 million years before the end-Eocene (~34 Ma) an impact cluster of at least three large asteroids, creating the *Popigai* (Fig. 5.8) (Siberia, D ~ 100 km; 35.7 + 0.2 Ma), *Chesapeake Bay* (off-shore Virginia, D ~ 85 km, 35.3 + 0.1 Ma) and *Mount Ashmore* (Timor Sea, End-Eocene, D > 50 km) (Fig. 5.3) (Glikson et al. 2010) resulted in ~10 % extinction of species. The cluster, superposed on a cooling trend at the end of the Eocene, was followed by a sharp decline in temperatures likely associated with the opening of the *Drake Passage* between West Antarctica and South America, leading to the formation of the circum-Antarctic current. This isolated the continent from warm tropical influences, allowing the formation of the Antarctic ice sheet and global cooling by near to 5 °C (DeConto and David 2003; Sijp and England 2004).

Extra-terrestrial bombardments and long-lasting volcanic episodes left destructive marks on ancient biospheres. Through human history the passage and explosion of small comets left an indelible impression on the human mind. Comets, such as Halley's comet were regarded as bad omens heralding catastrophes, such as the death of kings. A recent expression was the Heaven's Gate cult related to the appearance in 1997 of Comet Hale-Bopp.

According to Ward's (1994, 2009) 'Medea hypothesis', the origin of large mass extinctions of species and loss of biodiversity are in part driven by microbial activity and sedimentary residues of life. Similar interpretation has been advanced by Bradshaw and Brook (2009). Whereas the triggers for mass extinctions could arise from external forcing, in terms of the 'Medea hypothesis' the magnitude of a mass extinction events may be greatly enhanced by feedbacks from organic matter-rich sediments such as carbonaceous shale and limestone and from microbial activity in reducing environments. Examples include:

1. Large scale release of microbially metabolized methane, identified by its low $\delta^{13}C$ signature, as recorded at the ~55 Ma-old Paleocene-Eocene Thermal Maximum (PETM) (Zachos et al. 2001, 2008);

2. The Permian–Triassic mass extinction, which is interpreted in part due to oceanic anoxia induced by enrichment of the oceans in methane, hydrogen sulfide and other compounds (Ward 1994, 2009) associated with large scale basaltic volcanism (Siberian taps) and likely the *Araguinha* asteroid impact.

Bradshaw and Brook (2009), acknowledging the effect of external factors on terrestrial evolution, describe the 'ebb and flow of life on Earth along a thermodynamic spectrum' as intrinsic to the biosphere. According to these authors 'extinction is an inevitable part of evolution' and 'extinction is a part of natural selection'.

The authors proceed with analogies between individual life cycles, speciation and extinction, stating, for example: 'the causes of extinction can be thought of as equivalent to the different processes that lead to individual deaths within a population'. In terms of its unique anthropogenic origin, the Anthropocene (Crutzen and Stoermer 2000; Ruddiman 2005; Steffen et al. 2007) constitutes a prime example of biologically induced mass extinction. However, the search for common principles in the history of the rise and fall of species should not overlook the temporally and spatially unique character of each of the above cataclysms and its distinct effects on biological evolution.

References

Alvarez W (1986) Toward a theory of impact crises. Eos 67:649–658

Alvarez L, Alvarez W, Asaro F, Michel HV (1980) Extraterrestrial cause for the Cretaceous-Tertiary extinction. Science 208:1095–1108

Benton MJ (2003) When life nearly died: the greatest mass extinction of all time. Thames and Hudson, London, pp 336

Bradshaw CJA, Brook BW (2009) The cronus hypothesis: extinction as a necessary and dynamic balance to evolutionary diversification. J Cosmol 2:201--209

Convey-Morris S (2003) Life's solution: inevitable humans in a lonely universe. Cambridge University Press, Cambridge, p 486

Crutzen PJ, Stoermer EF (2000) The 'Anthropocene'. Global Change Newslett 41:12–13

DeConto RM, David P (2003) Rapid Cenozoic glaciation of Antarctica induced by declining atmospheric CO_2. Nature 421:245–249

French BM (1998) Traces of catastrophe—a handbook of shock metamorphic effects in terrestrial meteorite impact structures. Lunar Planet Sci Instit Contrib 954:120

Glikson AY (2005) Geochemical and isotopic signatures of Archaean to early Proterozoic extraterrestrial impact ejecta/fallout units. Aust J Earth Sci 52:785–799

Glikson AY (2008) Field evidence of *Eros*-scale asteroids and impact-forcing of Precambrian geodynamic episodes Kaapvaal (South Africa) and Pilbara (Western Australia) Cratons. Earth Planet Sci Lett 267:558–570

Glikson AY, Jablonski D, Westlake S (2010) Origin of the mount Ashmore structural dome west Bonaparte Basin Timor Sea. Aust J Earth Sci 57:411–430

Hildebrand AR, Penfield GT, Kring DA, Pilkington M, Camargo ZA, Jacobsen SB, Boynton WV (1991) A possible Cretaceous-Tertiary boundary impact crater on the Yucatan Peninsula, Mexico. Geology 19:867–871

Keller G (2005) Impacts volcanism and mass extinction: random coincidence or cause and effect? Aust J Earth Sci 52:725–757

Pope KO, Kieffer SW, Ames DE (2004) Empirical and theoretical comparisons of the Chicxulub and Sudbury impact structures. Meteor Planet Sci 39:97–116

Raup DM (1991) Extinction: bad genes or bad luck? WW. Norton and Co, pp 210 + xvii

Ruddiman WF (2005) Plows plagues and petroleum: how humans took control of climate. Princeton University Press, Princeton, p 224

Sepkoski JJ (1996) Patterns of phanerozoic extinction: a perspective from global data bases. In: Walliser OH (ed) Global events and event stratigraphy. Springer, Berlin, pp 35–52

Sijp W, England MH (2004) Effect of the drake passage through-flow on global climate. J Phys Ocean 34:1254–1266

Stanley SM (1987) Extinctions. Scientific Am Library, New York

Steffen W, Crutzen PJ, McNeill JR (2007) The Anthropocene: are humans now overwhelming the great forces of nature? Ambio 36:614–621

Ward PD (1994) The end of evolution: on mass extinctions and the preservation of biodiversity. Bantam, New York, p 301

Ward PD (2007) Under a green sky: global warming, the mass extinctions of the past, and what they can tell us about our future. Harper Colins Publishers, New York

Ward PD (2009) The Medea hypothesis: is life on Earth ultimately self-destructive?. Princeton University Press, Princeton, pp 208

Watson AJ (2008) Implications of an anthropic model of evolution for emergence of complex life and intelligence. Astrobiology 8:175–185

Zachos J, Pagani M, Sloan L, Thomas E, Billups K (2001) Trends rhythms and aberrations in global climate 65 Ma to present. Science 292:686–693

Zachos J, Dickens GR, Zeebe RE (2008) An early Cenozoic perspective on greenhouse warming and carbon-cycle dynamics. Nature 451:279–283

Chapter 13
Uniformitarian Models and the Role of Asteroid Impacts in Earth Evolution

Abstract The progressive identification of impact ejecta units in early green-stone belts, testifying to ongoing bombardment during the mid-Archaean and late-Archaean, raises questions regarding uniformitarian and plate tectonic models of early crustal evolution.

Impact scientists have long suspected that large asteroid impacts have had a profound, hitherto unrecognized, effect on the history of the Earth's crust (Dietz 1964; Grieve 1980; Jones 1987; Glikson 1993, 1994, 1996, 1999, 2001, 2005a, b, 2007, 2008; Boslough et al. 1994). Several authors pointed out circumstantial evidence for temporal relations between large asteroid impacts, major faulting and onset of plate-tectonic events, including stages in the breakup of Gondwana (Alt et al. 1988; Oberbeck et al. 1992). These include correlations between large impact clusters, hot spots, continental basalts (Morgan 1981; Richards et al. 1989; Courtillot et al. 1999), continental rifting and opening of oceanic gaps. Given the Geo-centric focus of Earth science, to date only limited account has been taken of the effects of large asteroid impacts (Windley 1977; Condie 1995; Myers 1995; Smithies et al. 2005; Van Kranendonk et al. 2007).

Attempts to correlate Phanerozoic magmatic, rifting and plate-tectonic events with large asteroid impacts (Hughes et al. 1977; Jones 1987; Alt et al. 1988; Marvin 1990; Oberbeck et al. 1992) based solely on age overlaps remain subject for tests by further isotopic and stratigraphic age correlations. Possible candidates for such tests include (Table 1.2): (a) Late Triassic asteroid bombardment (Manicouagan ~214 Ma; D ~ 100 km); Rochechouart (~213 Ma; D ~ 23 km), succeeded by rifting of the Atlantic Ocean at ~200–180 Ma; (b) Late Jurassic asteroid bombardment (Morokweng 145 + 0.8 Ma; D ~ 70 km); Gosses Bluff (142.5 + 0.8 Ma; D ~ 24 km); Mjolnir (143 + 2.6 Ma; D ~ 40 km) succeeded by stages of Gondwana breakup; (c) End-Cretaceous impacts (Chicxulub 64.98 + 0.05 Ma; D ~ 150 km); Boltysh (65.17 + 0.64 Ma; D ~ 25 km), succeeded by opening of the Indian–Arabian gulf.

Isotopic age determinations of thermal events and peak crust formation episodes through the geological record, initially by Rb–Sr isochron ages (Moorbath 1975, 1977) and subsequently by whole-grain and ion-probe U–Pb zircon studies,

A. Y. Glikson, *The Asteroid Impact Connection of Planetary Evolution*,
SpringerBriefs in Earth Sciences, DOI: 10.1007/978-94-007-6328-9_13,
© The Author(s) 2013

including studies of igneous and detrital zircons (Compston et al. 1986; Compston and Kroner 1988; Nelson 2008; O'Reilly et al. 2008; Valley 2008), have progressively defined isotopic age peaks (Fig. 10.1). This raises the question regarding the origin of the corresponding thermal events, including the extent to which asteroid impact events may be reflected by the data. Isotopic age studies identify rocks crystallized or metamorphosed as early as about ~4.0 Ga, including the Acasta Gneiss, Slave Province (Iizuka et al. 2007) and in Antarctica (Harley and Kelly 2007). These terrains include banded granite gneiss derived from metamorphosed igneous and sedimentary formations, as in Greenland, Labrador, Slave and Superior Provinces of the Canadian Shield, Finland, South Africa, India, Western Australia and Brazil. Original models regarded these rocks as a primordial granitoid-dominated (*SIAL*) basement above which younger volcanics and sediments were laid (Hunter 1970, 1974; Sutton 1971; Oversby 1975; Bridgewater and Collerson 1976; Moorbath 1977; Windley 1977; Chadwick et al. 1978; Hickman 1981). This theory gained support from the discovery of pre-volcanic zircons in greenstone belts (Compston et al. 1986) and of detrital zircons up to 4.4 Ga in the Narryer Terrain of the Yilgarn Craton, Western Australia (Harrison et al. 2005). The paradigm of a primordial granitoid basement was challenged by Canadian geologists (Folinsbee et al. 1968; Goodwin 1974; Card 1990) who regarded andesite volcanics and sedimentary enclaves within Archaean gneiss as analogues of circum-Pacific island arc–trench systems. Geochemical evidence and isotopic age and initial $\delta^{143}Nd$ and $\delta^{176}Hf$ data (McCulloch and Bennett 1994; Kamber 2007; Pietranik et al. 2008; Valley 2008) are consistent with derivation of granitoid crust from mafic parental materials (Naqvi 1976; Tarney et al. 1976; Glikson 1972, 1984, 1999; Card 1990; Myers 1995; Kroner et al. 1996). Plate-tectonic-based Archaean crustal models dominate the Archaean literature, for example on the basis of eclogite xenolith and mantle depletion studies (James and Fouch 2002), whereas other authors regard rifts, marginal basins and volcanic rifted margins as the close modern analogues of Archaean basalt–komatiite–rhyolite-dominated greenstone successions (Bleeker 2002).

The philosophical paradigm of 'the present is the key to the past' (Lyell 1830) has not taken asteroid impacts into account. The question as to the extent to which early Earth processes and environments differed from modern plate-tectonic regimes is progressively resolved through the ongoing discovery of large impact ejecta units in Archaean terrains (Lowe and Byerly 2010). Significant differences exist between Archaean greenstone belts and circum-Pacific ophiolite–turbidite accretion wedges (Glikson 1980; Hamilton 1998). Engel (1966), impressed by documentation of the 3.5–3.2 Ga Onverwacht Group, Barberton greenstone belt, interpreted the sequence of pillowed Mg-rich quench basalts, peridotitic lavas and intrusive dolerite and gabbro (Viljoen and Viljoen 1971; Anhaeusser 1973), in terms of ancient oceanic crust. Trace-element studies (Sun and Nesbitt 1978) showed the Mg-rich lavas imply high degrees of melting of the early mantle and thereby high geothermal gradients. The common occurrence of supracrustal volcanic and sedimentary enclaves within granitoid gneiss and the U–Pb and Sm–Nd evidence for granitoid precursors of some of these sediments, resulted in a 'chicken and eggs' impasse. Some of the deformed mafic-ultramafic enclaves contain isotopic $^{207}Pb/^{206}Pb$ and $\delta^{143}Nd$ values suggesting pre-existence of yet older granitoids. An example is furnished by zircons entrained in supracrustal units of the Kalgoorlie greenstone belts, Western Australia (Compston et al. 1986).

Differences between ancient and modern environments include the vertical accumulation in greenstone belts of more than 10 km-thick volcanics and sediments over time spans as long as <300 Ma, for example from about ~3.5 to ~3.2 Ga in the Pilbara Craton (Figs. 3.2, 3.3) and Kaapvaal Craton (Fig. 3.3) (Poujol et al. 2003; Hickman 2004; Hickman and Van Kranendonk 2004; Van Kranendonk et al. 2007). Significant structural and lithological contrasts exist between Archaean greenstone belts and ophiolite–turbidite accretionary wedges in circum-Pacific margins and arc–trench chains, which accumulated over shorter time intervals (Hamilton 1998). These differences include the unique dome structure of granite–greenstone terrains, referred to by Macgregor (1952) as 'gregarious batholiths.' Isotopic age studies indicate Archaean granite–greenstone systems represent multiple episodes, each including both volcanic and plutonic components, contrasted with the relatively more continuous accretion of ophiolite–turbidite wedges around the circum-Pacific margin.

Green (1972, 1981) suggested that, due to high geothermal gradients on the early Earth, higher buoyancy of the oceanic crust constrained transformation of basalt into garnet-bearing eclogite, thus retarding circum Pacific-like gravity-driven subduction of oceanic crust. A variation of this concept is the low-angle subduction model of Smithies et al. (2003). Mantle-plume models (Griffiths and Campbell 1991; Hill 1991; Davies 1999; Campbell and Davies 2006) have not to date been established for modern volcanic systems. Mantle-plume models for the origin of Archaean mafic–ultramafic volcanism (Smithies et al. 2005; Van Kranendonk et al. 2007) do not discriminate between magmas derived by melting of convective upper mantle, deep upwelling of mantle plumes, and adiabatic melting of asthenosphere generated due to by deep faulting.

Few current models of crustal evolution have taken the tectono-thermal role of large asteroid impacts into account. By contrast, new discoveries of asteroid impact ejecta units in early Archaean terrains suggest an extension of the Late Heavy Bombardment, originally defined at 3.95–3.85 Ga (Ryder 1990, 1991, 1997), into the early Archaean (Lowe and Byerly 2010). The sharp compositional contrasts between units underlying and overlying impact fallout beds (Figs. 3.3, 8.3) militates for potential relationships between large impacts, deep crust–lithosphere faults, and onset of volcanic episodes enriching the oceans in soluble ferrous iron. The evidence indicated above (Sect. 8.1.2) for likely relationships between the 3.26–3.24 Ga-old multiple impacts, related unconformities and plutonic events in the Barberton greenstone belt and contemporaneous unconformities and mega-breccia units in the Pilbara Craton is illustrated in Figs. 8.4 and 13.1. By analogy to the transformation from an ultramafic–mafic crust at 3.26–3.24 Ga in close relation with impacts by an asteroid cluster, the concentration of large asteroid impacts toward the end-Archaean about ~2.63–2.48 Ga may provide at least a large part of the explanation for the Archaean-early Proterozoic transformation.

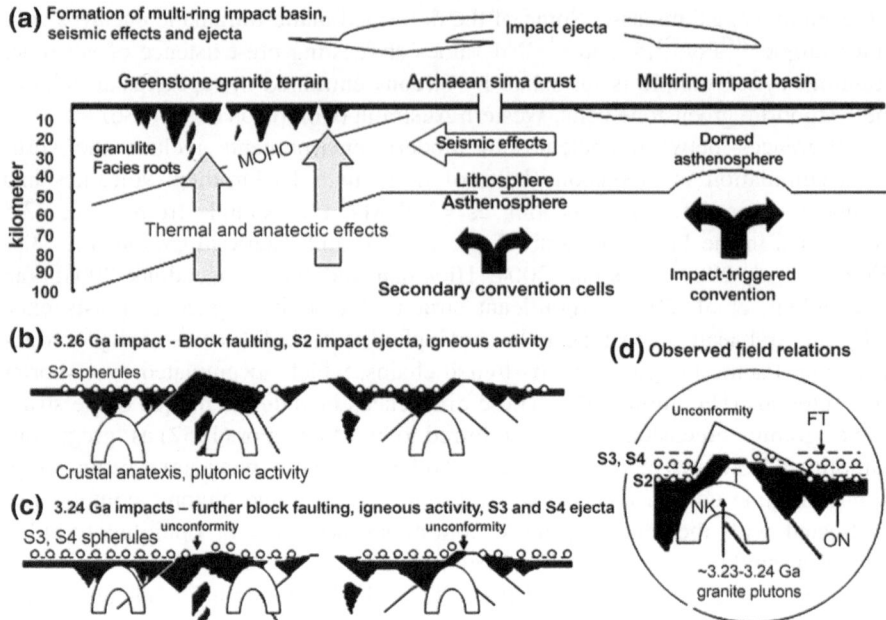

Fig. 13.1 A model portraying the principal stages in multiple ~3.26 and ~3.24 Ga impacts in an oceanic region of the Archaean Earth and their geodynamic consequences, including formation of oceanic impact Mare, mantle rebound and volcanic activity, ensuing rearrangement of mantle convection patterns, seismic activity affecting pre-existing granite-greenstone SIAL nuclei (micro-continents), faulting, uplift, erosion and formation of unconformities, anatexis at the roots of SIAL nuclei and rise of granitoid magmas. **a** ~3.26 Ga: formation of a multi-ring impact basin by a ~20 km asteroid, seismically triggered faulting, mantle rebound and onset of a new convection cell, thermal and anatectic effects across the asthenosphere–lithosphere boundary below sial nuclei. **b** ~3.26 Ga: Block faulting in SIAL nuclei, rise of anatectic granites, settling of S2 ejecta spherules, preservation of S2 spherules in below-wave-base environments. **b** ~3.24 Ga: S3 and S4 impacts, ejecta fallout and preservation in below-wave-base environments, further faulting, block movements and rise of plutonic magmas. **d** Schematic representation of observed field relationships between the ~3.55 to 3.26 Ga mafic–ultramafic volcanic Onverwacht Group (*ON*), intrusive early tonalite and trondhjemite (*T*), 3.26–3.24 Ga granites (*NK*), S2 ejecta, unconformity, S3, S4 ejecta, and the Fig Tree Group sediments (*FT*). (Glikson 2008; Elsevier, by permission)

References

Alt AD, Sears JW, Hyndman DW (1988) Terrestrial mare: the origins of large basalt plateaus hotspot tracks and spreading ridges. J Geol 96:647–662

Anhaeusser CR (1973) The evolution of the early Precambrian crust of southern Africa. Phil Trans Roy Soc London A273:359–388

Bleeker W (2002) Archaean tectonics: a review with illustrations from the Slave Craton. In: The early Earth: physical chemical and biological development, Fowler CMR (ed). Geol Soc. London Spec Publ 199:151–181

Boslough MB, Chael EP, Trucano TG, Kipp PP, Crawford DA (1994) Axial focusing of impact energy in the Earth's interior: proof-of-principle tests of a new hypothesis. In: new developments regarding the KT event and other catastrophes in earth history. Lunar Planet Sci Instit Contrib 825:14–16

Bridgewater D, Collerson KD (1976) The major petrological and geochemical characters of the 3600 my Uivak gneisses from Labrador. Contrib Mineral Petrol 54:43–56

Campbell IH, Davies GF (2006) Do mantle plumes exist? Episodes 29:162–168

Card KD (1990) A review of the superior province of the Canadian Shield—a product of Archaean accretion. Precam Res 48:99–156

Chadwick B, Ramakrishna M, Viswanatha MN, Murthy VS (1978) Structural studies in the Archaean Sargur and Dharwar supracrustal rocks of the Karnataka Craton. J Geol Soc India 29:531–542

Compston W, Kroner A (1988) Multiple zircon growth within early Archaean tonalitic gneiss from the ancient gneiss complex Swaziland. Earth Planet Sci Lett 87:13–28

Compston W, Williams IS, Campbell IH, Gresham JJ (1986) Zircon xenocrysts from the Kambalda volcanics: age constraints and direct evidence of an older continental crust below the Kambalda–Norseman greenstones. Earth Planet Sci Lett 76:299–311

Condie KC (1995) Episodic ages of greenstone: a key to mantle dynamics? Geophys Res Lett 22:2215–2218

Courtillot V, Jaupart C, Manughetti IT, Tapponnier P, Besse J (1999) On causal links between flood basalts and continental breakup. Earth Planet Sci Lett 166:177–196

Davies GF (1999) Dynamic Earth: plates plumes and mantle convection. Cambridge University Press, Cambridge 472 p

Dietz RS (1964) Sudbury structure as an astroblemes. J Geol 72:412–434

Engel AEJ (1966) The Barberton Mountain Land: clues to the differentiation of the Earth, vol 27. University of Witwatersrand Information Circle, South Africa

Folinsbee RE, Baadsgaard H, Cumming GL, Green DC (1968) A very ancient island arc. Am Geophys Union Monogr 12:441–448

Glikson AY (1972) Early Precambrian evidence of a primitive ocean crust and island nuclei of sodic granite. Geol Soc Am Bull 83:3323–3344

Glikson AY (1980) Uniformitarian assumptions plate tectonics and the Precambrian Earth. In: Kroner A (ed) Precambrian plate tectonics, Elsevier, Amsterdam, pp 91–104

Glikson AY (1984) Significance of early Archaean mafic–ultramafic xenolith patterns. In: Kroner A, Goodwin AM, Hanson GN (eds) Archaean geochemistry. Springer, Berlin, pp 263–280

Glikson AY (1993) Asteroids and early Precambrian crustal evolution. Earth Sci Rev 35:285–319

Glikson AY (1994) Archaean spherule beds: impact or terrestrial origin? Earth Planet Sci Lett 26:493–496

Glikson AY (1996) Mega-impacts and mantle melting episodes: tests of possible correlations. AGSO J Aust Geol Geophys 16:587–608

Glikson AY (1999) Oceanic mega-impacts and crustal evolution. Geology 27:341–387

Glikson AY (2001) The astronomical connection of terrestrial evolution crustal effects of post-3.8 Ga mega-impact clusters and evidence for major 3.2 Ga bombardment of the Earth–Moon system. J Geodynamics 32:205–229

Glikson AY (2005a) Geochemical and isotopic signatures of Archaean to early Proterozoic extra-terrestrial impact ejecta/fallout units. Aust J Earth Sci 52:785–799

Glikson AY (2005b) Geochemical signatures of Archaean to early Proterozoic mare-scale oceanic impact basins. Geology 133:125–128

Glikson AY (2007) Early Archaean asteroid impacts on Earth: stratigraphic and isotopic age correlations and possible geodynamic consequences In: Van Kranendonk MJ, Smithies H, Bennett VC (eds) Earth's oldest rocks. Developments in Precambrian geology, vol 15. Elsevier, Amsterdam, pp 1087–1103

Glikson AY (2008) Field evidence of *Eros*-scale asteroids and impact-forcing of Precambrian geodynamic episodes Kaapvaal (South Africa) and Pilbara (Western Australia) Cratons. Earth Planet Sci Lett 267:558–570

Goodwin AM (1974) Precambrian belts plumes and shield development. Am J Sci 274:987–1028

Green DH (1972) Archaean greenstone belts may include terrestrial equivalents of lunar Mare? Earth Planet Sci Lett 15:263–270

Green DH (1981) Petrogenesis of Archaean ultramafic magmas and implications for Archaean tectonics. In: Kroner A (ed) Precambrian plate tectonics. Elsevier, Amsterdam, pp 469–489

Grieve RAF (1980) Impact bombardment and its role in proto-continental growth of the early Earth. Precamb Res 10:217–248

Griffiths RW, Campbell IH (1991) Interaction of mantle plume heads with the Earth's surface and onset of small-scale convection. J Geophys Res 96:18295–18310

Hamilton WB (1998) Archaean magmatism and deformation were not products of plate tectonics. Precamb Res 91:143–179

Harley SL, Kelley NM (2007) Ancient Antarctica: the Archaean of the East Antarctic shield In: Van Kranendonk MJ, Smithies RH, Bennett VC (eds) Earth's oldest rocks. Developments in Precambrian geology, vol 15. Elsevier Amsterdam, pp 149–186

Harrison TM, Blichert-Toft J, Muller W, Albarede F, Holdren P, Mojzsis SJ (2005) Heterogeneous Hadean hafnium: evidence of continental crust by 4.4–4.5 Ga. Science 310:1947–1950

Hickman AH (1981) Crustal evolution of the Pilbara Block Western Australia. Geol Soc Aust Sp Publ 7:57–69

Hickman AH (2004) Two contrasting granite–greenstone terrains in the Pilbara Craton Australia: evidence for vertical and horizontal tectonic regimes prior to 2900 Ma. Precam Res 131:153–172

Hickman AH, Van Kranendonk MJ (2004) Diapiric processes in the formation of the Archaean continental crust east Pilbara granite-greenstone terrain Australia. In: Eriksson PG, Altermann W, Nelson DR, Mueller WU, Catuneanu O (eds) The Precambrian Earth: tempos and events. Developments in Precambrian geology, vol 27. Elsevier, Amsterdam, pp 54–75

Hill RI (1991) Starting plumes and continental breakup. Earth Planet Sci Lett 104:398–416

Hughs HG, App FN, McGetchin TN (1977) Global seismic effects of basin-forming impacts. Physics Earth Planet Inter 15:251–263

Hunter DR (1970) The ancient gneiss complex in Swaziland. Trans Geol Soc South Afr 73:105–107

Hunter DR (1974) Crustal development in the Kaapvaal Craton: part 1—the Archaean. Precamb Res 1:259–294

Iizuka T, Komiya AT, Maruyama S (2007) The early Archaean Acasta gneiss complex: geological geochronological and isotopic studies and implications for early crustal evolution. In: Van Kranendonk MJ, Smithies RH, Bennett VC (eds) Earth's oldest rocks. Developments in Precambrian geology, vol 15. Elsevier, Amsterdam, pp 127–148

James DE, Fouch MJ (2002) Formation and evolution of Archaean cratons: insights from southern Africa. In: Fowler CMR (ed) The early earth: physical chemical and biological development. Geol Soc of London Sp Publ 199:91–103

Jones AG (1987) Are impact-generated lower crustal faults observable? Earth Planet Sci Lett 85:248–252

Kamber BS (2007) The enigma of the terrestrial protocrust: evidence for its former existence and importance of its complete disappearance In: Van Kranendonk MJ, Smithies RH, Bennett VC (eds) Earth's oldest rocks. Developments in Precambrian geology, vol 15. Elsevier, Amsterdam, pp 75–90

Kroner A, Hegner E, Wendt JI, Byerly GR (1996) The oldest part of the Barberton granitoid–greenstone terrain South African: evidence for crust formation between 3.5 and 3.7 Ga Precamb Res 78:105–124

Lowe DR, Byerly GR (2010) Did the LHB end not with a bang but with a whimper? 41st Lunar Planet Sci Conf 2563pdf

Lyell C (1830) The principles of geology, Vol 2. Murray, London

Macgregor AM (1952) Some milestones in the Precambrian of southern Rhodesia: anniversary address by the President. Proc Geol Soc South Afr IIV:xxvii–lxxiv

Marvin UB (1990) Impact and its revolutionary implications for geology. In: Sharpton VL, Ward PD (eds) Global catastrophes in Earth history: an interdisciplinary conference on impacts volcanism and mass mortality. Geol Soc of Am Sp Pap 247:147–154

McCulloch MT, Bennett VC (1994) Progressive growth of the Earth's continental crust and depleted mantle: geochemical constraints. Geochim et Cosmochim Acta 58:4717–4738

Moorbath S (1975) Evolution of Precambrian crust from strontium isotopic evidence. Nature 254:395–398

Moorbath S (1977) Ages isotopes and the evolution of the Precambrian continental crust. Chem Geol 20:151–187

Morgan WJ (1981) Hotspot tracks and the opening of the Atlantic and Indian oceans In: Emiliani E (ed) The sea, vol 7. Wiley Interscience, New York, pp 443–487

Myers JS (1995) The generation and assembly of an Archaean supercontinents: evidence from the Yilgarn craton Western Australia. Geol Soc London Sp Publ 95:1439–1454

Naqvi SM (1976) Physical-chemical conditions during the Archaean as indicated by Dharwar geochemistry. In: Windley BF (ed) Early history of the Earth. Wiley, London, pp 289–298

Nelson DR (2008) Geochronology of the Archaean of Australia In: De Laeter JR, Gleadow AJW, McDougall I (eds) Geochronology in Australia. Aust J Earth Sci 55:779–793

O'Reilly SY, Griffin WL, Pearson NJ, Jackson SE, Belousova EA, Alard O, Saeed A (2008) Taking the pulse of the Earth: linking crustal and mantle events In: De Laeter JR, Gleadow AJW, McDougall I (eds) Geochronology in Australia. Aust J Earth Sci 55:983–996

Oberbeck VR, Marshall JR, Aggarval H (1992) Impacts tillites and the breakdown of Gondwanaland. J Geol 101:1–19

Oversby VM (1975) Lead isotopic systematics and ages of Archaean acid intrusives in the Kalgoorlie-Norseman area Western Australia. Geochim et Cosmochim Acta 40:1107–1125

Pietranik AB, Hawkesworth CJ, Storey CD, Kemp TI, Sircombe KN, Whitehouse MJ, Bleeker RW (2008) Episodic mafic crust formation from 4.5 to 2.8 Ga: new evidence from detrital zircons Slave craton Canada. Geology 36:875–878

Poujol M, Robb LJ, Anhaeusser CR, Gericke B (2003) A review of the geochronological constraints on the evolution of the Kaapvaal Craton South Africa. Precamb Res 127:181–213

Richards MA, Duncan RA, Courtillot V (1989) Flood basalts and hot spot tracks: plume heads and tails. Science 246:103–107

Ryder G (1990) Lunar samples lunar accretion and the early bombardment of the Moon. Eos (Trans Am Geophys Union) 71:313–322

Ryder G (1991) Accretion and bombardment in the Earth–Moon system: the Lunar record. Lunar Planet Sci Instit Contrib 746:42–43

Ryder G (1997) Coincidence in the time of the Imbrium Basin impact and Apollo 15 Kreep volcanic series: impact induced melting? Lunar Planet Sci Instit Contrib 790:61–62

Smithies RH, Champion DC, Cassidy KF (2003) Formation of Earth's early Archaean continental crust. Precamb Res 127:89–101

Smithies RH, Van Kranendonk MJ, Champion DC (2005) It started with a plume: early Archaean basaltic proto-continental crust. Earth Planet Sci Lett 238:284–297

Sun SS, Nesbitt RW (1978) Petrogenesis of Archaean ultrabasic and basic volcanics: evidence from rare earth elements. Contrib Mineral Petrol 65:301–325

Sutton J (1971) Some developments in the crust. In: Glover J (ed) Proceedings of symposium on Archaean rocks, Perth. Geol Soc Austr Sp Publ 3:1–10

Tarney J, Dalziel IWD, DeWitt (1976) Marginal Basin 'Rocas Verdes' complex from south Chile: a model for Archaean greenstone belt formation. In: Windley BF (ed) The early history of the Earth. Wiley, New York, pp 131–146

Valley JW (2008) The origin of habitats. Geology 36:911–912

Van Kranendonk MJ, Smithies RH, Hickman AH, Champion DC (2007) Paleoarchaean development of a continental nucleus: the east Pilbara terrain of the Pilbara Craton Western

Australia. In: Van Kranendonk MJ, Smithies RH, Bennett VC (eds) Earth's oldest rocks. Developments in Precambrian geology, vol 15. Elsevier, Amsterdam, pp 307–338

Viljoen RP, Viljoen MJ (1971) The geological and geochemical evolution of the Onverwacht Group in the Barberton Mountain Land South Africa. In: Glover J (ed) Proceedings of symposium on Archaean rocks, Perth, Geol Soc of Australia Sp Publ 3:133–151

Windley BF (1977) The evolving continents. Wiley, London 385 p

Chapter 14
The Current Danger

Abstract Current observations of the asteroid flux in the Solar system suggest future impacts are inevitable and will continue to affect life on Earth, as they have in the past.

By 2012 the number of Near Earth Asteroids (NEA) detected by NASA reached 9,252, of which ~2,250 objects are of diameters ~100–300 m, ~2,800 objects of diameters 300–1,000 m and 850 objects of diameters >1 km (NASA 2012). Estimates suggest at least a thousand NEAs may be large enough, i.e. ~1.0 km or more, to threaten Earth. On June 30, 1908, a small asteroid 100 m in diameter exploded over Tunguska, Siberia, devastating more than half a million acres of forest. One of the most recent close calls occurred on March 23, 1989, when an asteroid 0.4 km large came within 640,000 km from Earth. The potential seismic and atmospheric effects of impact, by even a moderate-size asteroid of ~100 m, on human habitats, in particular cities, fuel and communication systems, render civilization vulnerable. Current estimates suggest some 25,000,000 asteroids of this order of magnitude in the solar system, while evidence for past impacts on the terrestrial planets, including Earth, is growing. The effects of large impacts and volcanic episodes on the history of life on Earth are summed up in Fig. 1.2.

The danger of potential future impacts has not been lost on NASA's scientists. When on the 12 February, 2001, the space probe NEAR Shoemaker, named after the renowned planetary geologist Eugene Shoemaker, rendezvous with *Eros*, the 34 × 11 × 11 km-large asteroid (see Frontispiece), the encounter signified the first time Earthlings made contact with an asteroid. In 1998 NASA instigated a Near Earth Object (NEO) search program aimed at identifying asteroids larger than 1 km in size. The Catalina Sky Survey (CSS) and its affiliated Siding Spring Survey (SSS) utilize three telescopes, a 60 inch telescope, a 27 inch telescope and a 20 inch telescope, located in Arizona and in Siding Springs, Australia. The CSS has become the most prolific NEO survey. Only occasionally are NEOs missed by these surveys.

NASA's Near-Earth Object Program monitors Earth-approaching and Earth-crossing asteroids of sizes down to a few meters, but mostly down to a few tens of

A. Y. Glikson, *The Asteroid Impact Connection of Planetary Evolution*,
SpringerBriefs in Earth Sciences, DOI: 10.1007/978-94-007-6328-9_14,
© The Author(s) 2013

meters. The history of recent asteroid near-Earth encounters, in terms of asteroid sizes and distance from Earth, includes the following:

- Asteroid 2003-UV11, D ~ 370–820 m; relative velocity 25.3 km/s, close approach at 5 lunar distances (1 LD = 384,000 km) on 30 October, 2010.
- Asteroid 1998-TU3, D ~ 3.3–7.4 km; relative velocity 9.85 km/s, approach at 69.1 Lunar Distances on 17 October, 2010.

The consequences of an impact by asteroid 2003-UV11 would depend on the location of the impact, incidence angle and nature of the target. An oceanic impact would result in major tsunami and, should the body hit land or reach the sea floor, in regional to global earthquakes, fires and dust clouding. A land impact by asteroid 1998 TU3 would lead to a regional atmospheric flash, igniting fires. It would create a crater about 50–100 km across, trigger high-magnitude earthquakes, cause global dusting leading to years of cooling (the so-called asteroid winter effect) and lead to release of greenhouse gases (water, CO_2, methane, nitric oxide) from target rocks. The long-lasting effects of the latter would raise mean global temperatures by several degrees for periods of centuries, millennia and longer.

On the 29 September, 2004, asteroid 4179-Toutatis, a 4.6 × 2.4 × 1.9 km-large body (Fig. 14.1), came as close to Earth as 4 Lunar Distances (1 LD = 384,000 km (Fig. 14.2). The next close approach is projected to occur on the 12 December, 2012, missing the Earth by distance ~7.10^6 km (18 LD). Toutatis is an Apollo type (Earth crossing) asteroid with a small orbit inclination (i = 0.47 arc.deg), hence a possibility to approach closely the Earth (Sitarski 1998). This author states "Our

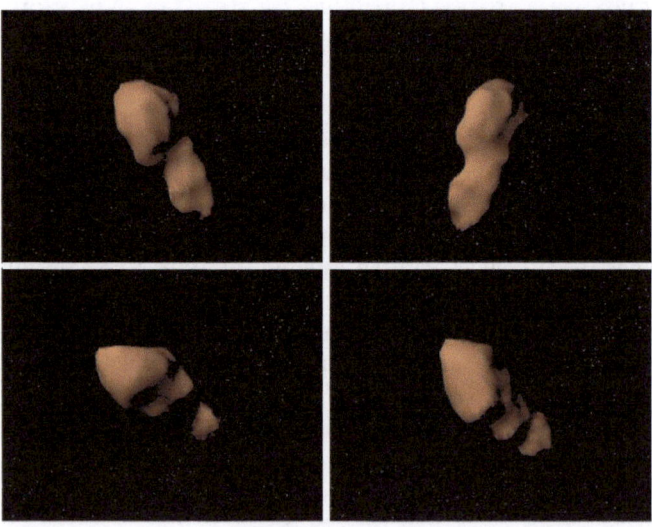

Asteroid 4179 Toutatis

Fig. 14.1 Images representing rotation of asteroid 4179 Toutatis. http://echo.jpl.nasa.gov/asteroids/4179_Toutatis/toutatis.html (NASA)

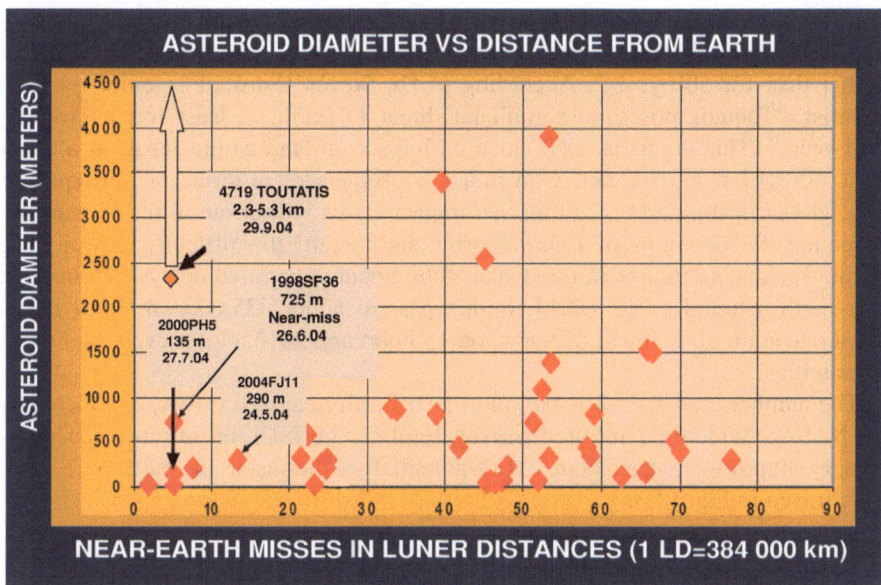

Fig. 14.2 Asteroid diameters vs distance from Earth (in Lunar distances) during 2004. Data source: NASA

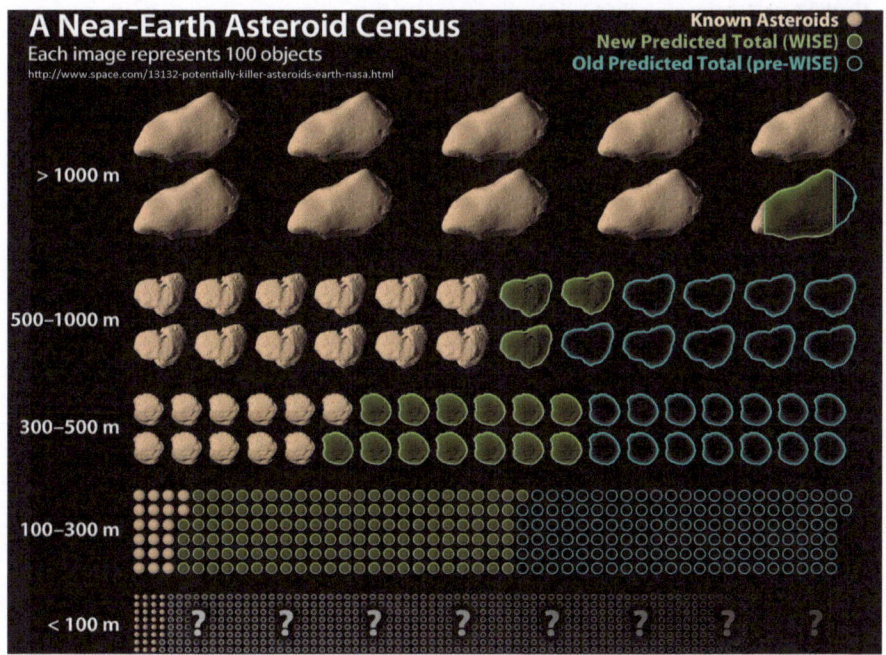

Fig. 14.3 Size distribution of near Earth Asteroids (NEA) conducted by the NASA's Wide-field Infrared Survey Explorer (WISE), an infrared space telescope (http://www.space.com/13132-potentially-killer-asteroids-earth-nasa.html). Credit: NASA

results confirm a conclusion found by other authors that Toutatis orbit is exceptionally chaotic. Therefore, we are not able to predict the motion of Toutatis further than for 300 years". According to Dr. Steven Ostro, JPL senior research scientist, "Toutatis poses no significant threat to Earth, at least for a few hundred years". This is just as well since its impact on land would result in a crater some 40–60 km in diameter, with major consequences in terms of Earthquakes, fires, global dusting and acid rain. An impact at sea will ensue in major tsunami. Given the chaotic nature of *Toutatis'* orbit, and thereby the difficulty in projecting its movements, an impact in the distant future cannot be ruled out. A ~45 m-large near-Earth asteroid, 2012 DA14, is due pass at about ~35,000 km from Earth, about one tenth of the lunar distance, on 15 February, 2013, closer even than orbiting satellites.

The numbers and size distribution of near Earth Asteroids (NEA) conducted by the NASA's Wide-field Infrared Survey Explorer (WISE), an infrared space telescope (http://www.space.com/13132-potentially-killer-asteroids-earth-nasa.html) is portrayed in Fig. 14.3.

Reference

Sitarski G (1998) Motion of the minor planet 4179 Toutatis: can we predict its collision with the Earth? Acta Astronomica 8:547–561

About the Author

 Andrew Glikson, an Earth and paleo-climate scientist, studied geology at the University of Jerusalem and graduated at the University of Western Australia in 1968. He conducted geological studies of the oldest geological formations in Australia, South Africa, India and Canada, studied large asteroid impacts, including effects on the atmosphere, oceans and mass extinction of species. Between 1998 and 2012 this work included detailed studies of impact ejecta units in the Pilbara Craton, Western Australia, geochemical studies of impact ejecta units from South Africa, and the identification and study of several Australian buried impact structures, including Woodleigh, Gnargoo, Mount Ashmore, Talundilly and Warburton. Since 2005 he extended the studies of past mass extinctions to the effects of climate on human evolution, the discovery of fire and global warming. He was active in communicating nuclear issues and climate change evidence to the public and parliamentarians through papers, lectures and conferences.

A. Y. Glikson, *The Asteroid Impact Connection of Planetary Evolution*,
SpringerBriefs in Earth Sciences, DOI: 10.1007/978-94-007-6328-9,
© The Author(s) 2013

Index

A. Y. Glikson, *The Asteroid Impact Connection of Planetary Evolution*,
SpringerBriefs in Earth Sciences, DOI: 10.1007/978-94-007-6328-9,
© The Author(s) 2013